TIME

BIG IDEAS

General editor: Lisa Appignanesi

As the twenty-first century moves through its tumultuous first decade, we need to think about our world afresh. It's time to revisit not only politics, but our passions and preoccupations, and our ways of seeing the world. The Big Ideas series challenges people who think about these subjects to think in public, where soundbites and polemics too often provide sound and fury but little light. These books stir debate and will continue to be important reading for years to come.

Other titles in the series include:

Julian Baggini **Complaint**
Jenny Diski **The Sixties**
Paul Ginsborg **Democracy**
Ian Hacking **Identity**
Steven Lukes **Moral Relativism**
Susie Orbach **Bodies**
Renata Salecl **Tyranny of Choice**
Slavoj Žižek **Violence**

TIME

Eva Hoffman

To Bea,
with thanks for a
lovely supper —

PROFILE BOOKS

Eva Hoffman
Sept. 17, 2009

First published in Great Britain in 2009 by
PROFILE BOOKS LTD
3A Exmouth House
Pine Street
London EC1R 0JH
www.profilebooks.com

1 3 5 7 9 10 8 6 4 2

Typeset in Minion by MacGuru Ltd
info@macguru.org.uk
Printed and bound in Italy by L.E.G.O. S.p.a. – Lavis

A CIP catalogue record for this book is available from the British Library.

ISBN 978 1 84668 038 0

CONTENTS

… How sour sweet music is
When time is broke and no proportion kept!
So is it in the music of men's lives.

Shakespeare, *Richard II*

But at least, if strength were granted me for long enough to accomplish my work, I should not fail … to describe men first and foremost as occupying a place, a very considerable place compared with the restricted one which is allotted to them in space, a place, on the contrary prolonged beyond measure … in the dimension of Time.

Marcel Proust, *Remembrance of Things Past*

Time present and time past
Are both perhaps present in time future,
And time future contained in time past …
What might have been and what has been
Point to one end, which is always present.

T. S. Eliot, *Four Quartets*, 'Burnt Norton'

INTRODUCTION

Let me begin with a confession: I have always been preoccupied by time. Whether this propensity was temperamental, or whether in some way it belonged to the place and historical moment in which I was growing up, I am not sure. But I do know that it started early, and has continued throughout my adult life.

Perhaps, like the young Vladimir Nabokov, I was simply a kind of chronophobiac.[1] Certainly, I was intensely and palpably aware of time's existence and its ceaseless passage. Reading some childhood story, in which the ticking of a clock measures the silent night, I would start listening to the clock in my own room, aware that each tick-tock was irreversible, and that the stealing of time, second by second, would never stop. When I walked home from school on some warm afternoon, I was conscious that with each step taken a moment was receding behind me into the past, that the number of such moments a life had in it was finite and that the only way to preserve them in some way was to hold them in my mind; in memory. I would remember this moment, I would say to myself; that way, it would not be entirely lost.

Children are natural philosophers; I suspect that more of them experience metaphysical sensations than we know. Nabokov became aware that he existed in the 'pure element' of time at the age of four, and he compared the birth of this consciousness to a 'second baptism, on more divine lines than the Greek Catholic ducking' seen performed on his younger brother. Perhaps my own exacerbated sense of time's unstoppable passage arose out of the climate in which my childhood took place. I grew up in Poland, shortly after the war – that is, on the territory of vast death. The presence of mortality was pervasive and inescapable there. We post-war children knew in our bones that life was a provisional condition; that it could be cut arbitrarily; that its finitude was wrested briefly from death's infinity.

These, to be sure, were powerful circumstances; but perhaps they suggest that our basic vision of time can be established quite early, and may be informed by the cultural and historical context we arrive into. But there was another kind of time I was aware of (for various temporalities can coexist, and be folded into each other as subatomic dimensions are apparently folded into each other in the cosmos). This was the ordinary time of our daily lives, and of the human activity all around us. That time moved at an unhurried, temperate, seemingly just-right pace. Of course, what I was experiencing was the unrushed time, the *tempo giusto* of childhood; but even adult time in Cracow during those years seemed to move more slowly than in any of the places I lived in since then. I don't think this was merely a lyrical illusion. Rather, as I reflect on it in retrospect, I see that this earlier, slower tempo was partly a function of the actual conditions of people's lives and partly a question of cultural ethos or temper. Poland in the post-war decades was an impoverished country and an economically static society. Under the aegis of centralised communism, there were no great careers to be made, no glamorous possibilities of upward mobility or the seductive temptations of acquiring great wealth. There was really nothing much to hurry towards. The famous motto of those days, 'We pretend to work, and they pretend to pay us,' summarised in a joke what was often a grim material reality, the downside of social stasis.

But the virtue of these defects was the sufficiency, the plenitude of time. People had time to sit around the famous eastern European kitchen table, and talk late into the night; people had time to ruminate inconsequentially during a slow amble in a park. Here's Carmen Firan, a Romanian poet now living in New York: 'For more than thirty years I lived in the opaque world of communism, where time had no value,' she writes. 'All we had left was talking. Our conversations, sometimes delightful, were a never-ending chatter over full ashtrays and cheap bottles of alcohol, night-long discussions, and hung-over mornings. Time was frozen for us. We weren't in a hurry to get anywhere. Neither did we have anywhere to go.'[2]

People had time for such things. But they also did not question their need or purposes, did not wonder about the returns or the results, or the intrinsic worth of sitting about and talking late into the night. For there was something else in the air of those days, something more elusive than social conditions, and more difficult to demonstrate or pin down: a predisposition to value purely personal and intimate experience, and to savour the textures of that experience; a predilection for a kind of pensiveness, for musing on small things, and reflecting on larger ones. In other words, a predilection for taking one's time about the flow of living.

This *was* the lyrical side of the famous Slavic melancholy. But such atmospheres, too, reflected deeper cultural patterns. It seems possible that, aside from the effects of communism, the slower time of eastern Europe was a feature of a less puritanical climate, of cultures which (for better and for worse) had never developed a full-blown capitalist ethos, or the idea that time is money. Indeed, the idea of excessive ambition, of running after things, seemed vaguely unseemly and undignified. 'When man is in a hurry, the devil makes merry', was an oft-cited (and here roughly translated) proverb of my childhood; and the devil, in those days, was still a personage to reckon with. The word 'fate' was often used; and the acknowledgement of that governing force in people's lives implied a certain receptivity, a willingness to accept things as they are, which we, in our more activist – or will-driven – societies might call passive, or find difficult to imagine.

Dislocation exacerbates the consciousness of time. For me, emigration constituted a great interruption, putting paid to the idea that time necessarily unfolds in a continuous, linear way. The past was all of a sudden on the other side of a great divide, preserved in memory but severed from the present. The future was so obscure and veiled as to have no existence. Straightforward temporal coordinates had become scrambled. Time, in a sense, had stopped flowing and started to assume more jagged forms and rhythms. My experience in this regard was, of course, hardly exceptional. Migrations, mass movements of populations as well as more routine forms of mobility, are among the hallmarks of our epoch and they inform everyone's consciousness of lived time. It is geographical stability and the continuous life narrative which these days constitute the exception, and various forms of discontinuity and fragmentation which are becoming the norm.

But for me, the impact of cultural disruption had more than personal implications. As I continued to live in America and study and work in its institutions, I began to become aware of the deep differences in the constructions of time prevailing between the two worlds I knew. It was not only that time moved faster in America – it pressed onwards in more stressful ways. People worked much harder, of course; but also, it seemed to me, more anxiously. I was witnessing, even if I did not initially realise it, the phenomenon of 'American nervousness' which had been a trope of social commentaries ever since the end of the nineteenth century. The nervousness had always been diagnosed as a function of a peculiar American insecurity, underlying the ostensible confidence; an uncertainty which followed perhaps partly from the country's perpetually renewing newness, but also from the extreme competitiveness of American institutions and the very possibilities of upward mobility. People worked very hard. But even if not everyone used every minute of their working day to be optimally productive – so I noted during my tenure in some major

American workplaces – everyone suffered from the stress of not doing enough, or the possibility of doing more, or at least feeling good and guilty about it. (This soon included myself.) It was as if anxiety were the tithe paid to the gods of the work ethic in lieu of more concrete sacrifices. After all, everything was at stake in American careers: big promotions, big money, big homes. And if you didn't succeed in 'making it', as the colloquial phrase had it, you had only yourself to blame.

All of this meant that, on the whole, people were much more strict in their management of time and much less willing to give it away freely, or indulge in an errant impulse. The spontaneous response to a friend's summons, the leisured conversation which did not fit into anything and led nowhere in particular, the protracted silence between people as they let some thought sink in, or simply sat side by side together, all of this was much less likely to occur in the American context. Time needed to be apportioned rationally, with an eye for its yields and gains – even if these were defined as 'productive' encounters or conversations.

The different attitudes towards time, I gradually realised, required a different organisation not only of one's Filofax (the period's time-management device of choice), but of the self and one's internal arrangements. I learned quite a few things from American time: the merits, in some areas of life, of efficiency and rigorous schedules; the need, in a busy and complicated life, to know the worth of one's hours, and to assert one's temporal rights; the pleasures of pitching the self towards a directed goal, with a plan and a will to carry it through. I continue to value such attitudes, and the particular kind of self-cultivation – the disciplines of self – they encourage. By the time I travelled through eastern Europe in the immediate aftermath of the collapse of Soviet communism in 1989, I was no longer willing to wait patiently in long queues, and I contemplated a roomful of Romanians, waiting impassively in a large and dusty room for some documents to be processed, with a kind of anthropological and slightly irritable wonder.

And yet: it often seemed to me that in the insistence on imposing control on time's shaping, and on optimal efficiency in most of life's areas, in the subordination of so many activities to the calculations of productivity, and to the pursuit of predetermined objectives, that in all of this keeping of the eye on the ball, something – or even much – was lost. What that something was is rather difficult to define. The costs of stress, the syndromes of Type-A personality (that popular archetype of psychological profiling, characterised by such traits as impatience, excessive time-consciousness, aggressive competitiveness and inability to relax), the heart attacks among middle-aged professional men, were all well recorded in post-1950s America. But it also seemed to me, as I watched (and imbibed) the routine tensions of American life, that what was being traded in was nothing more or less than the experience of experience itself. And what is that? Perhaps something like the capacity to enter into the textures or sensations of the moment; to relax enough so as to give oneself over to the rhythms of an episode or a personal encounter, to follow the thread of feeling or thought without knowing where it leads, or to pause long enough for reflection and contemplation. Carmen Firan again: 'Time is everything in America. It is sold at each deli and hot-dog cart, on TV and by insurance companies, on slot-machines or in the *Have a nice day* greeting everyone utters automatically only to get rid of you quickly ... Time is money. The Soul? It is lying lonely somewhere on a shrink's chair, in front of a computer screen or in a cell phone.'[3]

It takes time to penetrate another culture's sense of time. In my years living in Britain I began to discern yet a different sense and deployment of the temporal medium. The British character (if one can still speak of such an entity amidst the extremely rapid changes in that society's structures and composition) seems exceptionally activist but less exclusively focused on work than the American personality. The British have been great exponents of amateur endeavours, of lifelong learning, of far-flung travel. In informal conversations, observers who have lived in both countries often note, with considerable surprise, that British professionals seem to engage in a number of activities much greater than their American counterparts; and yet, the word 'workaholic' hardly fits here. A socioanthropological survey of the world's major cities revealed that people walk fastest in London. But this is combined with a projection of a much more relaxed, or at least nonchalant, personal style. Whereas Americans want to show their gods – or their bosses – that they are making heroic efforts, the British prefer to be seen as accomplishing their tasks with no effort at all, and preferably with some insouciance to spare. In a sense, theirs seems to be a more deliberative attitude towards time. They take their time deciding what to do, and set about whatever it may be with a will, but (at least ostensibly) without anxiety. The British stance, their refusal to be agitated or harried, suggests that they are the lords of time and that they'll be pressed by no man, or career incentive.

These are elusive matters, not easily pinned down, and undoubtedly not completely generalisable. What seems certain to anyone who has lived in different parts of the world, or even travelled extensively, is that cultural attitudes to time can have far-reaching implications for the ways we live, for forms of sensibility, and for the tenor and textures of experience. I suppose that, broadly speaking, what I had been observing in my own trajectory were the divergences between the cultures of fatalism, or acceptance; and those of will, or control. What sorts of difference such differences make is sometimes difficult to see from the outside; but in conversations with various eastern Europeans – most of them devoted to living fully intentional lives – a nostalgia for that earlier, more indulgent time sometimes surfaces. It is Milan Kundera, a Czech transplant to Paris, who has written an essayistic novel called *Slowness*, a sustained paean to languor and the sensuality of an unhurried, preindustrial pace. And on one occasion, a group of assorted Slavs among whom I found myself determined (but gave up after a luxuriously prolonged conversation) to write a 'Slow Time

Manifesto'. That said, the length of those post-war Polish vacations now seems to me not only indulgent but indolent. I could no longer tolerate that much rest, except under extreme pressure. I may have become temporally bicultural.

But in any case, such nostalgia may soon lose its object and address. The patterns of time have changed in eastern Europe, as they have everywhere – including America itself – since I first started observing, and feeling, the effects of the classical work ethic. Many people all over the world, even if they stay in their own countries, are experiencing the tremors and tensions of change as their societies leap from pre-modern rhythms and habits into more productive but more ruthless constructions of time. That was certainly one of the seismic shifts I sensed as I travelled through eastern Europe and the Balkans shortly after the transitions of 1989.

But beyond such changes, and sometimes perhaps obfuscated by them, there is the other kind of time; the one that filled Nabokov with fear and rebellion as he discovered that 'our existence is but a brief crack of light between two eternities of darkness'; the one by which I was troubled (in both senses of the word) as a post-war child. That more fundamental entity is both the great ineffable and everyone's metaphysical medium and element. We live in our bodies and psyches, in families, landscapes and nations; but above all, we live in time. It is the one dimension of experience we cannot leap out of, at least until the final act. But we can contemplate it, investigate it, get acquainted with its nature and workings. Indeed, the need for reflection, for making sense of our transient condition, is time's paradoxical gift to us, and possibly the best consolation for its ultimate power. Time gives us our existential premise, and coming to terms with it is equivalent to grappling with the great questions. And so, I wanted to press my fascination further, and to find out if one can become more intimate with time, to ask how it shapes our lives, and what may be our happiest dealings with it; and also to discover – insofar as possible – what philosophical fortification may be gained against its invisible laws and inevitable passage.

~

This little book arose out of one chronophobiac's – and chronophiliac's – concern with time. But it was also prompted by the sense that the character of lived time is changing in radical and unprecedented ways; that temporal discomforts are widely expressed and felt and that the question of time, previously left to professional philosophers or Slav ruminators, has become a public issue as well as a private problem.

The problematisation of time can be felt in a number of ways. The pervasive sense of stress, rush and pressure in contemporary societies has become a subject of policy debates as well as of routine complaint. Amidst other forms of prosperity, and even as we live longer than ever before, we seem to be suffering from endemic shortages of time. For some, in the inversion of the old motto, time has become a more valuable and less attainable commodity than money. People sometimes work very hard in order to 'buy' themselves free time. We no longer only manage time, we hire time-management consultants to help us deal with this unruly element. If forms of neurosis mirror social conditions, it may be significant that in the last three or so decades we have seen the widespread identification of an entirely new pathology – attention deficit disorder – whose symptoms (or causes?) have to do, essentially, with the decrease of tolerance for sustained units of time.

But the changes are more than sociological or purely subjective. It is the shifts in the external, 'objective' conditions of time that are the most startling. The changes proceed from several sources: the simultaneous acceleration of time and attempts at its extension, from technological manipulations, and from broadly shifting patterns of activity. If time has been the great subject of science fiction, it is because we have assumed that, for all its psychic malleability, its actual workings can be reshaped only in fantasy and the imagination. But these days we are tampering with those workings in reality, and at a level so deep as to seem material. On the micro scale, we are compressing time into ever more condensed and minute segments. Computer time, which is increasingly the medium in which we live, functions in nanoseconds and is making hitherto unimaginable speeds concrete. It is habituating us to ever faster and shorter units of thought and perception, and to a focus on the immediate present.

At the other end of the scale, we are experimenting

with methods for extending longevity which are stretching the boundaries of the human lifespan in previously unthinkable ways. And, as we move through time with more speed and freedom, temporality becomes increasingly severed from the natural cycles of years, days and seasons. In jet travel we conflate night and day without regard to the twenty-four-hour cycle. We can, if we wish, repeat 'summer' or 'winter' several times a year. But even if we remain in one place we can function, via our communications technologies, simultaneously in every time zone on the globe. That means that any one instant contains our awareness of simultaneous events in different geographical and temporal locations. Previous complaints about the severing of time from nature had to do with the insertion of clock time into the cycles of days and seasons. But our cognition of time is no longer even linked to the time through which we physically move. Rather, our experience of temporality is becoming increasingly deterritorialised and virtual.

All of this amounts to a paradigm-shift comparable to the Copernican or Einsteinian revolutions. About twenty years ago Jeremy Rifkin, an American sociologist, wrote that 'Politics, long viewed as a spatial science, is now also about to be considered as a temporal art. The politics of territory is about to be joined by the politics of temporality.'[4] By this, Rifkin meant questions of work and free time, or the social effects of fast-rising technologies. The latter have progressed by quantum leaps since then. But what is at stake is not only the public sphere, but the politics of experience. Our transactions with time affect us as crucially as any of the more familiar forces of, say, ideology or identity, and they have profound consequences for the quality of our lives and the deep processes of subjectivity.

In a sense, time has been reality's last unconquered frontier. In the last decades we have seen announcements of the end of nature, end of history, and the human, as the beginning of the post-human has been declared. But are we also coming to 'the end of time', as we have known it? In writing this essay I wanted to try to understand the character of contemporary time, and the enormous changes we are witnessing in our dealings with that dimension of experience. I also wanted to ask what, in the light of latest research and observations, we have come to know of human time and its workings. What is the latest thinking about the biology and neurology of time? What do we understand about the pace and rhythms of subjectivity?

It is one of the fascinations of time that it is both the most intangible of entities and the most inexorable. On one level it is a malleable substance, shaped into different forms by both culture and psyche. But more fundamentally, it has always been taken as the great given, the category of reality which – outside fantasy or fable – cannot be changed, deconstructed, or wished away. Humans have always struggled with it, and cried out against it, but they have always assumed there's no getting away from its relentless passage, or from the governing fact it bestows upon us of our mortality.

And so I wanted to explore both the flexibility of

human time and its possible limits. Are there, outside of cultural and psychological constructions, certain temporal universals? Is there such a thing as natural human time, or at least temporal norms, beyond which we venture at our peril? Are we, in trying to stretch temporal boundaries, liberating ourselves from age-old constraints, or striking a Faustian bargain at a potentially ruinous cost?

The time revolution is taking place not in the interplanetary sphere but within our daily lives, perceptions, even bodies, and with enormous repercussions for social relations, forms of experience and, indeed, for our very understanding of what it means to be human. The changes may be unstoppable, and in some ways even desirable. But they need to be reckoned with and grasped if we are to understand their possible consequences, and our current condition in the world.

1

TIME AND THE BODY

Is there such a thing as specifically 'human time', and if so, what is it? Perhaps it is possible to delineate a very broad framework for human temporality by noting what it most obviously is not. The human mind is capable of conceiving, and even measuring, scales and dimensions of time which clearly do not belong to the sphere of human or even earthly possibilities. We can calculate – even if we cannot exactly imagine – intergalactic aeons and the speed of light; we can catch light from extinguished stars whose radiance has taken thousands of light years to reach us, and we can capture sub-microscopic fractions of seconds which we have no way to experience.

It is surely a remarkable feature of the human mind that it can imagine such vastnesses and infinitesimal smallness; but these extremities do not belong to the temporal scale of our own experience. That scale is defined by some basic factors which so far have been unchangeable: the fact that we are organic, and have a particular biological make-up, and the equally determining fact that we live in this particular corner of the cosmos, and on this particular planet – our own familiar earth.

Before being human, we are living creatures, and this creates the first temporal condition of our existence. In the cosmic spaces, time 'in itself' can apparently move in various directions, including backwards. But for us it moves in only one direction: from the past to the present, from growth to decay, and from birth to death. This, reassuringly enough, may itself have good cosmological reasons. In *A Brief History of Time*, Stephen Hawking speculates that 'complicated organisms' like ourselves can only exist in a particular kind of universe – one which happens to be very much like ours.[1] In other words, we are adapted to our universe – or the universe to us – in what Hawking calls the 'strong version' of the 'anthropic principle'. One of the adaptations is to the direction of time. The universe we are in works on the principle of entropy, or the movement from order to disorder, and from initial mass to expansion and decay. This is why, in our kind of universe, in Hawking's view, the 'psychological arrow of time' is pointed in the same direction as the cosmological and the thermodynamic arrow of time (the order in which entropy increases): from the past to the future. Intelligent life, Hawking posits – that is, the kind of life which would have a perception of 'past' and 'future' – is possible only in a universe in which the arrow of time moves in this direction. In addition, as many cosmologists have noted, by the standards of intergalactic spaces, conditions on our own planet are characterised by a highly exceptional moderation (at least, moderation by our standards). The earth is, comparatively, temperate in size, temperature and, indeed, temporality – that is, the length of its revolutions around the sun, and therefore its days and nights.

It is the length of that daily cycle which is the first

terrestrial determinant of human temporality, as it is of other living creatures'. Whether Hawking's hypothesis about life's adaptation to our corner of the universe proves to be true or not, recent studies have made it clear that we are, on the biological level, quite precisely adjusted to the specific rhythms and cycles of the earth.

In the last few decades, problems of biological time have given rise to a new scientific discipline called chronobiology, within which various features of temporal behaviour are investigated through increasingly microscopic observations and precise measurements. One of the fundamental findings to emerge from these studies is that, within their appointed lifespan – and, so to speak, lifestyle – all biological species living on planet earth are adapted to earthly time, and its day/night cycle. This is one of those facts that can be taken for granted, or that can strike us as quite wondrous. For, whether animals are diurnal or nocturnal, their behavioural patterns follow, with some small adjustments, the twenty-four-hour cycle (as we have come to measure it). Moreover, the adjustment to the diurnal cycle does not occur through anything as simple as sight, which would let an animal know if it is day or night, or through any other sensory signals from the environment. Rather, the periodicity of living creatures is governed endogenously, from within, by versions of a marvellous mechanism called the 'biological clock'.

The facts which have been unearthed about these biological devices, and the seeming ability of non-percipient creatures to 'tell time', are endlessly fascinating. There is, for example, the famous dance of the bees, through which bees returning from feeding tell their fellow workers in their dark hive where the nectar is located. They do so by dint of performing a series of movements called 'waggles', which specify distances of the feeding site from the hive, the directional coordinates and the best time, in relation to the position of the sun, to arrive. And the bees do arrive, with uncanny accuracy, at the optimal time for the nectar to be collected.

How this happens, how biological clocks are constructed and the intricate processes involved in their operations, are still a matter of scientific speculation and research. However, while the actual biorhythms vary widely from species to species (some sleep most of the time; some hunt at night), it seems that the principles on which biological clocks operate are remarkably similar throughout the animal spectrum, suggesting that they have evolved from a single initial mechanism, or genetic code, rather than through convergent evolution. Internal clocks are among the oldest features of living organisms, to be found even in the most primitive bacteria.

Moreover, while the rhythms established by the biological clocks vary widely among species, it is another wondrous fact that in all species, and at every scale of organic functioning – genetic, metabolic, behavioural – these internal pacemakers work on the same basic principle, that of oscillation. This is a kind of movement, which, like the swinging pendulum, advances along a certain path only to be restrained and pushed back in the opposite direction. For example the rate of metabolism and bodily temperature oscillates regularly in many species, falling to a certain level before an internal feedback mechanism tells it to start climbing again. (Parenthetically, it is interesting to note that oscillation is the basic method of measuring time in man-made clocks as well; without the principle of regular motion, a division into regular units, both external and internal time would be an indiscriminate flow.)

The motion of oscillation, then, introduces temporality in living organisms, and is its basic pulse, or measure. Indeed, it can perhaps be said that internal temporal organisation is a fundamental requirement, as well as symptom, of life. Trying to answer the question posed by the physicist Erwin Schrodinger, which was simply 'What is life?', a mathematician, Jonathan D. H. Smith, posits that 'biological systems' are 'systems complex enough to isolate their component space-times' – in other words, to structure the indeterminate flow of time and space through internal dynamics.[2] Organisms are highly organised forms, and the internal structuring of time through regular motion seems to be among the earliest and most universal features of organic life. This, in fact, corresponds to what we actually know through simple observation: living organisms are entities which move of their own accord, while inanimate matter moves only under the application of external force. The measurement of organic time through oscillation also echoes philosophical intuitions dating back to Aristotle, who thought that time is the measure of motion ('the numeration of continuous movement').

But while movement seems to be a basic index of life, the Buddhists, in their contemplative quest for serenity, may also be expressing a physiological wisdom – for it seems to be a tendency of all organisms to seek homeostasis, or state of balance. This is one effect of introducing negative feedback into the oscillations, and not allowing the motion to progress too far in one direction. For example, if an animal's temperature rises too much, reverse hormonal or metabolic processes kick in to cool things down.

The internal motions of oscillation have their unique pace in each species, and in each the pacing is coordinated at various levels of functioning – genetic, metabolic and circadian. In higher animals this happens through intricate feedback mechanisms, so that, for example, hormones and neurons work in concert, producing regular behaviours through the diurnal cycle. In mammals the main 'clock' determining temporal behaviour has been found in a small cluster of paired cells located within the hypothalamus area of the brain known as the suprachiasmatic nuclei. (Interestingly, this is exactly the spot to which Buddhists point as the focus of contemplation, and which they call the 'third eye', because it is supposed to open in a state of enlightenment.) Aside from this central mechanism, mammalian rhythmic patterns are encoded in a number of genes – including the per or 'periodic' gene – working in complex interactions with each other and transmitting their information not only into the brain, but throughout the organism.

Moreover, there seem to be correlations not only

among various internal processes, but between different scales of species' temporality. In mammals, for example, the rate of heartbeat seems to be inversely proportional to lifespan. Elephants live seven times longer than mice, and an elephant's heartbeat is seven times slower than a mouse's. Does that mean that mice, within their brief span of days, 'feel' that they live as long – that they have the same amount of 'experience' – as elephants? Is the perception of experiential quantities related to the speed of internal processes? As far as mice and elephants are concerned, we will probably never know for sure, although from the human point of view the possible correlation of longevity to metabolic rhythms is suggestive. People with higher metabolic rates do seem to have shorter lifespans; this is one index which accounts for the differences in longevity between men and women. Men have shorter lives and higher rates of metabolism; but although subjective testimonies on this score might be easier to acquire for men than for mice, they are hard to find. Still, it is doubtful whether most men would feel that 'living harder' compensates for living less long.

In addition to the daily cycles, some species are adapted to annual or even longer-term rhythms. Migrating birds make their treks regularly at the same time of the year; cicadas, rather amazingly, emerge from below ground to lay their eggs at either thirteen- or seventeen-year intervals, almost to the day. This suggests that, even if animals may not be aware of the future – in our sense of 'awareness' – some instinct of futurity, or long-term temporality, seems to be built into many organisms. It is surely such an instinct – or mechanism of inner temporal regulation – which allows squirrels to put away food for the winter, or which drives certain fish to time their mating so that their offspring can emerge in the most favourable seasonal conditions.

Does that mean that some animals, in some sense, 'know' about the future – and what sort of 'knowing' would this be? The question of animal perception of time, as of animal experience altogether, remains a philosophical conundrum of the same order as that posed by the philosopher Thomas Nagel in his famous essay, 'What Is It Like to Be a Bat?'[3] What is certain is that each species has its specific time-patterns, or its temporal nature, which remains, within certain parameters, quite fixed, and which changes only through the extremely slow processes of evolution. Each animal is, in a sense, adapted not only to its own environment but, so to speak, to itself. Why and how that comes about seems to be, so far, less clear. One might cite Antony, in *Antony and Cleopatra*, as he tries to answer the question about what kind of thing is the crocodile: 'It is shaped, sir, like itself,' Antony answers, 'and it is as broad as it has breadth. It is just so high as it is, and moves with its organs. It lives by that which nourisheth it, and the elements once out of it, it transmigrates.' And, Antony might have added, it progresses through its days and nights very much at its own pace.

~

Can the same be said of humans? Are our temporal parameters also biologically calibrated? Do we, like Antony's crocodile, have a nature which is right and proper to us? Certainly, like other living organisms, we have a set of temporal principles built into our very genes and cells. Like other animals, we tick on biological clocks and, as in other species, these work on the principle of oscillation and are adjusted to the diurnal cycle. On the most basic level, time comes into existence from within our bodies, and we experience its tempos physiologically: through the rhythms of our heartbeat, or the alternations of appetite and satiation, or the pace at which we walk.

But the question of temporal nature is, of course, much more complicated for humans than for other animals – for humankind, uniquely, is the species which is not only susceptible to evolutionary or environmental change but which causes change, intentionally as well as accidentally. Throughout our history we have manipulated all aspects of our environment and, to some extent, our own selves – and temporality has been no exception. The human lifespan has consistently increased since *Homo sapiens* was first spotted on the horizon. At various times and places people have worked seasonally or all year round, have spent most of their time in energetic activity or not doing anything much. Unlike migrating birds, we can follow our caprice and take off for parts unknown at any season. Mechanical transport speeds up the pace of our passage through time; meditation slows it down.

Still, are there limits to our biological flexibility? And

what happens when we try to step outside them? That question is today being more dramatically tested, and contested, than ever before, and in some areas it still awaits complete answers. However, there are certain things we do know. We know, for example, that if our heartbeat speeds up or slows down beyond a certain point, we get heart attacks or die. We know that humans cannot run much faster than the four-minute mile. Seventeen seconds have been shaved off the record since that first sub-four-minute run by Roger Bannister in 1954, but unless we inject superdrugs or hyper-charged microchips into our bodies – unless, in other words, we become bionic, rather than human as we understand that term – a two-minute mile is highly unlikely.

We also know that humans need sleep. Sleep is a complex and seemingly highly flexible temporal behaviour. In earlier times, people adjusted their sleep to sunset and sunrise; they went to bed shortly after sunset, and slept longer hours in the winter than in summer. In medieval European villages, peasants often got up in the middle of the night and engaged in various activities before going back to (the usually crowded) bed. The optimal amount of sleep varies from person to person, and metabolic temperaments are divided between early-birds and night-owls. And yet, as studies of the sleep state and its disorders proliferate, it is increasingly clear that each person needs his or her metabolically conditioned quota of it in order to function at optimum capacity, or even to function at all.

On one level, sleep is the organism's way of conserving

energy and refuelling its supplies. All mammals need intervals of sleep, and in all of them, sleep patterns are regulated by the same circadian rhythms and homeostatic mechanisms which affect every aspect of our physiology. That is, the drive for sleep increases with the length of time that an individual has been awake – and vice versa.

But sleep involves an alteration not only in metabolism but in the state of consciousness – although what exactly that is, or what it is for, remains to some extent a puzzle. Sleep is hardly the same as unconsciousness. Things happen while we sleep, some of them of considerable importance. There has been speculation that one crucial purpose of sleep states is to enable the brain to store new information into long-term memory or, to put it technically, to transfer data from the motor cortex to the temporal lobe. This is accomplished through a recently discovered phenomenon called 'sleep spindles' – one- or two-second bursts of brain waves that intensify and subside at strong frequencies. These episodes of sudden activity occur during so-called REM (rapid eye movement) sleep, when we are most likely to dream. It is during REM sleep that the brain replenishes certain chemicals, known as neurotransmitters, which establish the connective neural networks essential for such crucial mental and physical functions as remembering, learning, problem solving and performance. And it is during REM sleep, and the bursts of sleep spindles, that new data we have taken in during the day is laid down more firmly in the neural structures of the brain, and thus transformed into knowledge which becomes more permanently a part of us.

Such descriptions of sleep processes are interesting, among other reasons, for their analogies with psychoanalytic speculations about dreams and their purpose. Freud famously declared that 'Dreams are a royal road to the unconscious' and it has been one of the tenets of depth-psychology since then that the purpose of dreams is to gain the kind of access to the hidden mainsprings of our motives and feelings which we do not have in waking life – and thereby to gain new perspectives on our conflicts and desires. Is it possible that, on the physiological level, this is literally what happens? That the brain, through the reorganisation of neural patterns, reorganises our understanding of our own experience as well? And that the information we have absorbed is embedded and integrated into the deeper, less conscious parts of neural memory? It has been found, for example, that if we learn a new skill (golf, playing the violin), and don't sleep afterwards, the benefits of the learning vanish. If we get sufficient sleep, the new skill apparently becomes encoded in our brain and body, so that we can perform it more automatically, and build on it the next time. It seems a small step from there to suppose that this may be true of more personal, emotionally charged information as well.

Whatever the exact links between neurology and psychology turn out to be, it is clear that sleep is crucial to our physiological as well as our cognitive and affective well-being. Conversely, it is statistically evident that chronic sleep deprivation can take a serious toll on our physical and mental health. Several aspects of the immune system are maintained by sleep, and without it they begin to weaken. The sense of mental wobbliness and anxiety which follows a sleepless night also has a physiological basis. Psychological stress and sleep deprivation both manifest themselves in excessive or diminished output of certain substances in the body – most notably the hormone cortisol. When it is produced at normal levels, cortisol can act as an anti-inflammatory, anti-stress substance. When it is produced in excessive quantities, however, as it is in conditions of sleep deprivation, cortisol suppresses the immune system and can contribute to many illnesses, including cancer.

There have been, throughout history and literature, famous insomniacs and heroic non-sleepers: Macbeth, who 'murders sleep' through his crime; Proust, who stays awake quiveringly in his bed until his mother comes in to give him his goodnight kiss; Nabokov, who declared sleep to be 'the most moronic fraternity', because it reduces everyone to an egalitarian non-sentience. In another vein, both Napoleon and George Sand apparently lived their hypercharged lives on four hours of sleep a night, and it is said that Margaret Thatcher, in her Iron Lady days, followed a similarly Spartan regime.

But in our work- and efficiency-driven societies, sleep disturbances are no longer a purely personal idiosyncrasy – a symptom, say, of exceptional vitality, or an exquisitely fine-tuned sensibility. Rather, sleep deprivation has become both endemic and systemic, a widely discussed public issue with its own institutions, statistics, ideologies and improving ideas. According to a poll conducted by the National Sleep Foundation in 2005, 'fewer than half of Americans say they get a good night's sleep every night, or almost every night', and the trends in Britain and other advanced countries are similar. Addressing such syndromes and symptoms are various research institutes, medical centres where people are taught how to sleep, and, in the US, the National Sleep Institute, the American Academy of Sleep Medicine and the Better Sleep Council. It seems that, for many people, this most seemingly natural of functions is now something that has to be learned.

Some of the causes of endemic sleep shortage have to do with the material conditions or requirements of our lives. Shift work, frequent long-distance travel, the availability of electric light and the incessant activity of cities all contribute to people getting less sleep. The work routines of upscale professionals often call for greatly extended or irregular hours. American medical students in their first year of hospital internship until recently worked seventy to eighty hours a week in shifts of up to thirty-two hours. In Britain a proposal to reduce the workload of junior doctors from fifty-six to forty-eight hours a week is planned, although the present fifty-six-hour maximum is probably exceeded regularly in practice. Stockbrokers and financial advisers in, say, New York have to be awake during not only their own working day but those in London and Tokyo as well.

But perhaps the deeper clues to the prevalence of

sleep pathologies are to be found in our attitudes to time. The hours spent sleeping are an interval in which we cannot 'do' anything, and nothing is seemingly accomplished; this, in our societies, is seen as either inefficient, vaguely shameful or, at the very least, a sheer waste of time. Perhaps just as saliently, sleep entails the temporary yielding of active will, and subsiding into a passive state, and this too may cause difficulty within cultures which value staying in control. 'The most damaging and persistent delusion we've acquired about sleep,' writes Jon Mooallem, 'is that this vital human function is optional. As one psychologist puts it, "You don't have people walking around figuring out how to get by on less air."'[4]

But sleep is not optional. It is a biological necessity, and its lack has socially measurable and sometimes dire costs. The medical interns working their protracted hours showed a 16.2 per cent increased risk of being involved in a road accident during their post-work commute, as well as diminished attention and a rise in diagnostic mistakes during overnight work in intensive care units.[5] Tired pilots are more likely to fly poorly. Athletes perform less well after a sleepless night.

Moreover, no matter how severed from the natural cycles of day and night we have tried to become, sleep continues to be stubbornly attuned to them. This is partly shown by statistics on shift work, indicating that people who work at night suffer from diminished attention and increased stress as well as a markedly greater incidence of ulcers, heart disease and even cancer. You might think that getting the required amount of sleep during the day would be as good as getting your REM-rest during the night. But this is not so, and the reason has to do with those internal clocks. However we try to fool ourselves that artificial light is as good as daylight, the circadian clocks within us continue to behave as though they know better. For example our body temperature is lower at certain times of day than at others, independently of what lights are on or off. So is the production of certain hormones and chemicals. One substance which contributes to the creation of sleep states, and through which the body is aligned to diurnal cycles, is melatonin. In its natural form, melatonin is the principal hormone of the pineal gland, and it is released into our bodies only at night. Its levels begin to rise at dusk and fall at dawn, in anticipation of our awakening, and its production is severely inhibited in the daytime. The shift workers, labouring by artificial light at night and trying to create night in the daytime, may simply not get enough melatonin for good, restorative sleep.

In its synthesised form, melatonin is used to counteract jet-lag – that is to adjust our bodies to the abrupt and arhythmical changes in time zones through which they are transported. But jet-lag itself is another phenomenon which suggests that we try to prise ourselves out of our natural diurnal cycles at our peril. Even if we sleep during our flight, and even if we expend no energy as we sit in our aeroplane seat, our bodies and brains fall into fatigue and disorientation as they are pulled out of alignment with their familiar sequence of day and night. Sometimes – to speak more impressionistically – the state of jet-lag seems to induce almost literal inner oscillations, as if perhaps our bodies, thrown off their familiar rhythms, were trying to refind their homeostatic equilibrium. More concretely, sports teams travelling long distances find that their performance suffers. Jet-setting executives may find that their negotiating skills at the other end of the trip are impaired.

As with terrestrial flights, so with the rather more rare but informative experience of space travel. Leaving the earth involves not only moving from one time zone to another, but entering zones where the day/night cycles come in different lengths. The days in a space shuttle orbiting the earth at speeds greater than its own revolutions are 23.5 hours long; a flight to Mars would result in a 24.65-hour day. Even such small variations in day length, together with exposure to the dim light of space shuttles, can lead to serious sleep loss and consequently decreased performance. The American space agency, for one, is taking the effects of such circadian disruptions on astronauts very seriously, and seeking remedies which, incidentally, might have applications in other circumstances.

Aside from the effects of sleep deprivation on our bodies and passing mental states, there are also evident links between disruptions of sleep and more persistent psychological problems or physical illnesses. The precedence of the chicken or the egg in such correlations is not always clear; we do not know whether certain mental states lead to sleep pathologies or vice versa. But what can be observed is that depression, anxiety and certain kinds of phobias are all characterised by severe distortions of sleep patterns: insomnia on the one hand, excessive sleep on the other. It is possible that the physiological link between these phenomena can be found in certain neuropeptides – substances secreted in various areas of the brain which both reduce the capacity for sleep and are associated with anxiety and depression.

Excessive need for sleep can be a sign of inner discomposure – depression, hormonal imbalance, or even simple unhappiness – as much as chronic insomnia. We do not want to sleep our lives away, like Rip van Winkle, or to spend them in the hypnotic state of Tennyson's 'Lotos-Eaters' – a poem whose vision of perpetual serenity is actually eerily disturbing ('All things have rest, and ripen toward the grave / In silence – ripen, fall, and cease: / Give us long rest or death, dark death or dreamful ease').

But in our societies it is sleep deprivation rather than 'dreamful ease' which is the characteristic disorder, and an index of certain cultural tendencies and values. The British writer Jeanette Winterson, in her short story 'Disappearance I', imagined a dystopia where sleep has been declared illegal – although some people are paid to fall into this state, so that others can observe their dreams.[6] We haven't gone quite that far; but perhaps more than in any other previous society, sleep, for us, has become a problematic site of morality and even a kind of Puritanism – if not yet illegal, then faintly illicit.

Such attitudes, however, are breeding, in the next swing of the psychocultural pendulum, their own sometimes highly ironic reactions. Apparently, in addition to not sleeping enough, many people are now anxious *about* not getting enough sleep – the proper number of hours for this state being determined by various statistical means and providing a new index of biological correctness. The anxiety about falling asleep in turn creates a kind of vicious-cycle recipe for insomnia, and has given fodder to a huge 'sleep industry' whose products guarantee getting good sleep in the most efficient manner. Some American corporations, noting that sleep deprivation costs US businesses billions of dollars in absenteeism and lost productivity, are now requiring their employees to take half-hour 'power naps' in the afternoon. 'Sleep for success' is a slogan of one sleep-industry brochure.

In such directives, sleep has somehow been reconfigured not as a natural function but as an activity – something we must accomplish. But this, of course, is to subordinate sleep to the same ethos of efficiency which has given rise to sleep problems in the first place.

Moreover, the new biological correctness doesn't dispense with the older moralism. However much we might want to fulfil our proper sleep quota, few would admit to sleeping long hours, or taking afternoon naps. Instead, we resort to drugs on a mass scale in order to sleep – and stay awake – more efficiently. Helpful companies throw up any number of over-the-counter pills and potions, especially popular among the young, ranging from megavitamins to energy drinks such as Red Bull to help us stay on the go. The use of such elixirs is certainly not the same as following natural cycles and predispositions. Among the more chemically enhanced substances, sleeping pills are well known to have damaging and sometimes frightening side effects, including anxiety and depression, and sometimes induce psychotic states. Amphetamines or cocaine, used routinely by whole rafts of high-flying professionals, may keep us artificially alert and speedy in thought; but if taken consistently, they make us ill and mad – or, as popular lingo aptly states, 'wasted'.

We want to get the most out of time, and to many of us this means being 'up' – in a literal as well as metaphoric sense – as much of the time as possible. And yet, time cannot be so easily cheated or subverted. The need for sleep is one of the indications that temporal patterning is deeply encoded into our biology, and that we try to violate it at high risk to ourselves. If we try to gain time by stretching it to the limit, then we may lose even more of it through physical illness or mental malaise. If we try to deploy our bodies and minds beyond their natural efficiency, we become less efficient. 'I wasted time, and now doth time waste me,' laments Shakespeare's Richard II. But it may be that, in our overdrive societies, we need to learn how to 'waste' time to some extent, or at least to accept the limits of our own temporality. If we do not, then time – in one of those inevitable paradoxes which it is given to producing – can waste us even more grievously.

~

The need for sleep is an index of our adjustment to

the earth's twenty-four-hour day/night cycle. Is there a similar coding for the *longue durée* of human life? Is there a normative lifespan for humans, as there is for other species? Are our days numbered by some long-acting internal clock, and is the pace of our life-stages governed by some preset measures as well? These are more complicated questions than ones concerning circadian rhythms, in part because lifespan is determined by such a multitude of factors and also because – as a result of being so over-determined – human longevity has varied widely, both historically and across different geographies and cultures.

The general direction of the longevity graph through recorded history, however, has been decidedly upward. When the ancients talked about the brevity of life, they really meant it. In Rome, at the height of its imperial power, the average lifespan was twenty-five years – although that included very high infant mortality rates. Once someone made it past the danger threshold of infanthood, they could expect to live forty years. In the intervening centuries there have been great setbacks, such as the Black Death, but the average lifespan has been steadily, if sometimes incrementally, lengthening. Since the onset of modern technology and medicine, the advances we have made in this area have been quite stunning. Global life expectancy is estimated to have risen from the average of forty-six years in the 1950s, to sixty-five years between 2000 and 2005, and this figure is expected to keep on rising, to seventy-five years in 2045–50. Within this broad statistical common denominator, however, there are wide discrepancies. In the least developed countries, life expectancy today is under fifty years, although it is expected to reach sixty-six years as we approach 2050. In the developed world, the average life expectancy for women these days is over seventy-nine years, and for men about seventy-four (although the gap between the sexes is closing). Apparently, by the middle of the twenty-first century, between 40 and 50 per cent of the population in the Western world and the Pacific Rim will be over sixty years old, with life expectancy at the upper end rising as well. One of the more startling statistics bred by the studies of longevity indicates that centenarians are the fastest growing age group in the US and Europe.

If we consider the sheer amount of time we are allotted for our lives to be a fundamental good, then clearly discrepancies in life expectancy constitute one of most fundamental inequalities between rich and poor parts of the world. The reasons for the disparities are fairly obvious; they have to do with availability of food and clean water, access to sanitation, medicine, inoculation against certain diseases and the means to protect newborn babies – infant mortality still accounting for a large portion of the differences in overall longevity rates. Occasionally there are culturally caused reversals of the general trend. In some eastern European countries after 1989, and especially in post-perestroika Russia, mortality rates for middle-aged men have risen dramatically; the sudden introduction of competitiveness in eastern Europe, alcohol and the sense of uselessness in Russia, have been adduced as reasons. But these are exceptions to the rule – interesting for what they reveal about the confluence of social factors and health, but not easily addressed by hygiene or technology.

Where possible, however, redressing inequalities by bringing life expectancy in the less developed world up to the standards obtaining in the most prosperous parts is a self-evident desideratum. But can we keep on increasing absolute longevity indefinitely? Is there an upper limit to the human lifespan – or can we, with sufficient knowledge and technology, perhaps live … forever? Until very recently, no one would have dreamed of asking such a question seriously; the very notion of endless life would have seemed either frivolous or dangerously Faustian in its hubris. Outside the fantasies of myth or science fiction – the biblical Methuselah and his fellow ancients of the prophetic age; the quests for the Fountain of Youth, or fables of alchemical transformation – humans have always accepted the facts of senescence and decline, and their natural, if not easily acceptable, terminus. Various beliefs in the afterlife notwithstanding, people have always observed that humans, like other animals, have their appointed lifespan, and that this is measured, roughly – and unlike the existence of a butterfly, say, or an oak tree – in decades rather than in centuries or days: a standard (at least by our human lights) neither utterly ephemeral nor especially long. Three-score and ten has been one proverbial measure of the time given to us; beyond that common denominator lay great old age, and the Great Inevitable.

But these days, the commonplace measure has been well surpassed, and the seemingly bedrock givens of age and mortality are being challenged in ways which range from the silly and the lunatic to the scientific and the perfectly serious. In the culture at large, a sort of anti-ageing movement is afoot, the members of which are determined to show that getting older doesn't have to mean getting old; that it is possible to postpone decline long past former expectations, and to stay alive longer than ever seemed possible before. Some of the methods by which people hope to accomplish this can be considered perfectly natural: more exercise, better diet. Other measures seem more radical, or more desperate. Some anti-ageists declare themselves to be, quite simply, anti-death, and put their faith in self-prescribed vitamin cocktails – the modern equivalent of the alchemical formula – in the hopes of fending off mortality until a more certain cure for it is found. Others resort to what is called 'extreme caloric deprivation' (not starving, exactly, but coming close to it). This technique is probably better founded in evidence, since such deprivation is one of the few factors shown demonstrably to prolong life – although not, one might perhaps note, in its most desirable form.

On the further fringes of experimentation, or wish-fulfilment, are the attempts to defeat the passage of time through cryogenics, or freezing the body in liquid nitrate for future revival. (Or should we call it resurrection?) There are organisations which specialise in this service, and which hold conferences on what is called 'extreme life extension'. At least a dozen multimillionaires have left wills specifying that they should be frozen just before death, and have bequeathed money to themselves in 'personal revival funds', on the chance that scientific advances will make it possible for them at some point to be restored to their former selves. Indeed, a number of souls (or at least bodies) are resting in large cryogenic containers already, waiting for some kindly scientist of the future to take an interest in them, and thaw them into sentient state. And, if you cannot fund such costly propositions, but still wish to achieve immortality, perhaps by more affordable spiritual means, you can, for example, enrol in the Jhershierra Seven Levels School for Mastery and Immortality, whose teaching, according to one satisfied customer, follows in the great tradition of 'hidden knowledge'. You can also purchase Jhershierra's book, *The Resurrected Dead Now Immortal Live Among Us: A Manual for Immortality* (credit cards of course accepted; interactive comments welcomed on the Immortality Message Board).

On the wilder margins of theoretical speculation, at least one reputable physicist has suggested that we will all be eventually revived, and that the Christian myth of resurrection has been scientifically sound all along. In a highly eccentric and scientifically intricate book, *The Physics of Immortality*, Frank J. Tipler puts forth the Omega Point Theory, which, he says, 'is a testable physical theory for an omnipresent, omniscient, omnipotent God who will one day in the far future resurrect every single one of us to live forever'.[7] The God whom Tipler posits is not the anthropomorphic personage of monotheistic religions, but is composed of physical forces, and the workings of time itself – which, at some point, Tipler stipulates, will invert itself, and start rewinding the history of our universe, so that everything which has decayed and died will reconstitute itself as it once existed.

At what personal age, and in what state of mind or dress, we may be revived is undoubtedly a nice question. In any case, such an apocalyptic outcome might take some billions of years to come round, so perhaps it is best not to worry about it too much. But in the meantime, the terrestrial quest for life extension, and even biological time reversal, is on. Moreover, some of it is occurring not only on the remoter margins of the culture at large, but within the precincts of biological laboratories and legitimate research. In the last decades the fundamental givens of ageing and even death have become objects not only of extravagant conjecture but of sober scientific scrutiny. Respectable scientists as well as self-proclaimed prophets have been questioning what has always been taken for granted: how, exactly, 'ageing' happens and what, exactly, it is for. The more rigorous explorations of such questions have yielded fascinating results, for it seems that when the phenomenon of ageing is examined with sufficient precision, much about it becomes less obvious than we might intuitively surmise. For example it is impossible to catch the organism as a whole in the process of getting older. After all, most of our cells, excepting those in the heart and the brain, are replaced many times throughout our lives. Moreover, they are replaced at different rates and different times for various organs. Why, then, should those organs, or the whole organism, become damaged or depleted?

The best candidate for an explanation is, in fact, found at the cellular and, increasingly, chromosomal level. It is cells and chromosomes rather than organs or organisms which, it seems, eventually exhaust themselves. Most cells in our bodies replicate themselves through division, but there is apparently a predetermined limit to the number of times a cell can perform this operation. This is called the Hayflick limit, after Leonard Hayflick, who discovered it in 1965. Moreover, Hayflick also demonstrated something which had been contested before: that normal cells have a finite lifetime, even when kept in cultures outside the organism. That means that the ageing of cells is not dependent on other processes working their way through the body; rather, the decline of cells seems to happen according to its own inbuilt schedule. Once a cell has gone through the appointed number of divisions, it reaches what is known rather suggestively as replicative senescence. Moreover, it has recently been found that the depletion of cell vitality has parallels at a chromosomal level. Each chromosome has specialised fragments of DNA, called telomeres, at its end, but on each cell division, a small number of these DNA bits are lost, partly through the action of an enzyme called telomerase. So, the longer we live, the shorter our chromosomal telomeres become. This has important implications, since the snipped chromosomes in turn become prone to strange fusions and recombinations, giving rise to errors in the replication of cells and to damaging mutations.

One of the effects of such errors is to produce free radicals – molecules with odd numbers of electrons, which attach themselves to other materials, upsetting the chemical balances in the organism. (Hence the current fad for anti-oxidants, which react quickly with free radicals, quenching their need for the aberrant electrons.) In some cases the errors in cell replication cause the chromosomes to cease producing crucial proteins which repair DNA, thus weakening our immune systems and our organisms. It is the working of such processes in our bodies which eventually is felt as a subsidence of strength and vitality; in other words, as ageing. And it is the gradual accumulation of such processes at the cellular and chromosomal level which eventually – even though different organs age at different rates, and even if our livers are much more healthy than our stomachs, or vice versa – leads to a general loss of function, a kind of waning of the entire interconnected bodily system. Eventually, even in the absence of a specific illness, the overall damage is sufficient to bring about the so-far still inevitable end.

The gradual processes of decline hold even for the recently discovered embryonic stem cells, on which so many medical and other hopes are pinned. These cells, found mostly in embryos but also in other tissue, are unspecialised and undifferentiated – that is, they are always found in their original state. When placed in cultures, embryonic cells can indeed renew themselves for much longer periods than differentiated cells, through division. But even they eventually reach the end of their capacity for self-replication and, in effect, on their micro level, die.

It is, moreover, one of the metaphysical ironies built right into our biology that an over-survival of certain cells in our bodies is itself dangerous, and can lead, through the recombinant errors and spontaneous mutations, to the development of malignant formations, including cancer cells. And it is a further irony that cancer cells and certain germ-line cells can divide without limit, and without damaging their own telomere lengths. Indeed, in these cells, an enzyme known as telomerase, whose action usually damages the telomeres in the life-perpetuating cells, can actually elongate the chromosome endings.

It seems, then, that even when the phenomenon of ageing is deconstructed into its component elements, the processes of decline and death are found to be encoded into the roots of our biology. Putting it more philosophically, it is tempting to say that finitude is the intrinsic cost of life, and that vulnerability is a necessary correlative of vitality. The question then arises, do such processes have a predetermined length and pace? Is there a temporal frame within which human existence takes place, and beyond which we cannot stretch? Some investigators of such data believe that, while the average lifespan has increased dramatically in recent times, the upper limit for human longevity has remained constant over many centuries – and that it hovers somewhere around the surprising figure of 120 years. The mythical ancients of rumour and legend may have existed in some remote regions even in pre-modern periods; on the other hand, nobody has managed to live beyond the twelve-decade boundary – at least so far. Possibly, we will discover that this is the natural limit of human existence and that – unless we become bionic and post-human – the number of our days is bounded by definite and delimited temporal parameters.

Certainly this is true for all other living organisms. How the duration of each creature's life is determined, or how various species are designed so as to live out their specific number of months, or years, or in some cases hours, is still – as with so many aspects of biological time – a matter of some mystery. But all species have their maximum age, beyond which they cannot endure. Annual plants flower and die in a season, but wild blueberry bushes – the longest-living species known to us – persist for 13,000 years. Animals living in the wild rarely reach senescence in their eat-or-be-eaten world; but bowhead whales apparently are capable of living upwards of 200 years. Mayflies, on the other hand, survive but for minutes, and then their teleological purpose is apparently fulfilled.

The puzzle of biological lifespan is intimately related to the most fundamental enigma of all – that of death. That mystery has always been taken to be philosophical, but in our age of scientific investigation, questions about the causes of mortality have been pushed back to the biological level. For example, in their important book *The Biology of Death: Origins of Mortality*, André Klarsfeld and Frédéric Revah take up the project suggested by a French scientist, E. Morin, who opined that 'Death must be Copernicised' – that is, that we must revolutionise our thinking about it, by examining its relationship to life as boldly and sceptically as Copernicus examined the relationship of the earth to the sun.[8] As part of that undertaking, Klarsfeld and Revah rather startlingly ask what death is *for*. The answer they arrive at logically is the one we might arrive at – once the question is so daringly posed – intuitively. Death has evolutionary advantages, most saliently to make room for the great variety of life and the succession of generations – and, therefore, for the possibilities of evolutionary selection and development. According to the ruthless laws of evolution, species exist not for the sake of individuals, but in order to perpetuate themselves. Beyond that function, we lose our uses, and evolution ceases to be interested in us. August Weismann, a pioneer in the studies of heredity, put this rather vividly (if cruelly) when he said that most of the organism is an appendage to reproductive cells. The selfish gene is very selfish indeed.

Within this explanatory framework, the time of death in each species is determined by the time of reproduction. But that, in turn, raises further questions about the great variety of relationships we witness in nature between these two events. There are species, such as the Pacific salmon, which suffer sudden death once their reproductive task has been fulfilled. There are others, such as certain kinds of spiders, in which the act of mating is closely followed by the female devouring its partner. In yet others, the very act of birth implies death – as in certain marine worms, whose eggs burst the parental body as they emerge from it. On the other hand, there are also species, such as sea anemones and flatworms, which seem to be capable of regenerating themselves indefinitely; and some fish, amphibians and reptiles, which keep growing larger but do not grow correspondingly older – although eventually, they also age and die.[9]

In the various possible correlations between reproduction and death, humans seem to represent a compromise. In most circumstances we can live for quite a while after losing reproductive capacities; but we age more quickly in our later years. Cellular errors and mutations increase with age. It is as if, in evolutionary terms, we become less interesting once we lose our reproductive powers – but not, apparently, entirely dispensable. Our lifespan, even under fairly minimal conditions, is designated to continue for a considerable time after childbearing age. One explanation for this is that evolution has programmed us to make sure we reproduce, but has also built in an extra quotient of years to make sure we bring up our progeny, and have some time to spare as well.

It must be said that such ex-post-facto explanations sometimes sound suspiciously close to the proposition that 'whatever is, is right'; that, like Shakespeare's crocodile, we are the way we are, because that is the way we happen to be. That, perhaps, is a sign that, while we can observe certain biological processes with increasing precision, their causality and eschatology are not as yet completely understood. What is perfectly clear, however, is that the lives of most organic creatures are circumscribed by definite temporal brackets; that all species are born, reproduce, and (unless they are killed first) die in strict sequence, and on pretty strict schedules.

But in this respect, *Homo sapiens* has been the great exception. Human life histories have never been as inextricably tied to biological life cycles as they are in other species. From early on we have cultivated grain and domestic animals to supplement our food supplies; we have built shelters to protect ourselves from the environment, and thus improve our life chances. We have created marriage laws which affect the time and rate of childbirth; for example the child brides of various periods and religions were meant to start giving birth early, and thus compensate for the high rates of both infant mortality and death in general. And it is one of the characteristics of our epoch that the separation between biology and life narratives is ever more complete. A significant proportion of people in advanced countries proclaim their independence from the putative reproductive aims of evolution by deciding not to have children at all. Others wait to have them till a more advanced age than was ever possible before; in a few cases, in vitro fertilisation has been administered to women in their sixties.

There is no question, however, that reproductive schedules still affect women much more signally than men. The bodily markers of fertility – the onset of menstruation, pregnancy, menopause – mean that women remain more closely tied to biological temporality than men; that the biological clock ticks for them more loudly, and occasionally sounds worrying alarms. The greater dependence on natural cycles and sequences perhaps also means that a sense of temporal omnipotence is harder to maintain for women than for men, or that it is at least tempered by a more inescapable awareness of lifetime's natural limits.

Such differences constitute a basic temporal inequality between the two genders. But on another level time is the great equaliser. Both women and men age, and eventually die – on the average, men rather sooner than women. And these days we are challenging biological temporality for both sexes in truly radical ways. We are trying to extend the time available to us beyond previous boundaries, and to manipulate the sequencing of life's stages beyond previous possibilities. And, on the further horizons of research, we are experimenting not only with preventing but reversing time's ravages. Some of the new means of rejuvenation and organic restoration paradoxically draw on the deepest workings of our biology – though with effects which would have seemed, until recently, entirely imaginary. One cutting-edge area of research with potential for life extension has to do with those seemingly miraculous embryonic stem cells and their regenerative properties. Since these cells can be induced, either in cultures or through being injected into a part of an organism, to become 'expressed' into any organ or tissue, they can, in principle, be used to repair – or eventually replace – failing parts of the body or physiological systems, from heart muscle to the liver, from insulin-producing cells of the pancreas to dopamine-producing glands. On the more futuristic frontiers, some laboratories are experimenting with transgenic transplants of organs from one species to another. This is a method whereby tissue from, say, a human heart, is implanted into the living heart of a pig, where it can develop into a heart viable for return into humans. Possibly, however, such stages of transgenic incubation might be unnecessary in the future, for already specific organs – a nose, an ear, a liver – can be grown directly in cultures from just a few cells with a little bio-chemical encouragement.

Such research is still in its experimental stages, and is subject to various legal constraints; but in some countries there are already perfectly legal bio-banks which can store a baby's birth-fresh stem cells, extracted from the placenta and the umbilical cord, for later use. Once such babies become adults, and begin ageing or falling ill, they will have the possibility of opening the safe and drawing on their very own biological materials to repair failing parts of the body, and perhaps even brain.

Those who can afford to invest in such promises of future rejuvenation might be forgiven for thinking that the Fountain of Youth has been finally discovered, and that its sources – mythically, allegorically – are to be found right within us, at the roots and origins of our biological being. In our own present it is rumoured that Russian plutocrats, and undoubtedly the mega-wealthy of other nationalities as well, receive illicit stem-cell injections not necessarily to fight illnesses, but purely for the purposes of feeling younger and better. And apparently, such injections can in fact have astonishing effects, such as smoothing out wrinkled skin or restoring waning potency and vigour. Whether the reparative or regenerative effects of stem cells can last permanently, or even over a long time, remains to be seen. But in the meantime, the Russian plutocrats could be excused for feeling that if the bargain they have struck with fate is Faustian, it is pretty shrewd nevertheless.

And then, of course, there are the already plausible prospects of cloning, with their accompanying promise of a sort of second life, or at least biological continuation – through the manufacture of a genetically identical avatar – beyond one's own extinction. A pioneering pet has already been cloned by an owner who couldn't bear its loss. Can human children be far behind? But whether we go on to produce humans by mechanical replication or stop short of it, the very knowledge that cloning and other body-replacement techniques are achievable alters our very conception of what it is to be human, and its most fundamental theorems: that personhood implies having one unrepeatable self, and living within the constraints of one unrepeatable body with a single irreversible lifespan.

Even short of cloning, the recent biological developments, and what they suggest about the possibilities of taking the production and continuation of life into our own hands, deeply change our relationship to time and all its workings. More and more we try to defy our temporal limitations and manipulate the patterns and stages of biological time. And, given the conscious and self-altering creatures we are, the urge of rebellion is understandable, even natural. Who would not want to postpone ageing, given a chance to do so? Who would not want to get a few extra years, or decades, if they could retain their faculties and good health with the prolongation? Old age has rarely been welcomed with open arms. There have been some traditional cultures in which the old have been respected, or even revered, for their accumulated wisdom: the patriarchs of the Old Testament for example, gain stature and power as they age. That must sweeten the pill of ageing considerably. But more frequently society has treated old people as dispensable and disposable. Indian widows in the poorer castes are even today relegated to humiliating subservience as they age. In times of scarcity the Eskimos used to expel the old from their igloos – after their teeth had fallen out and they could no longer chew their food – and leave them literally out in the cold, to die. Such practices were a function of survival and subsistence conditions, but even in less brutal times we have associated senescence with pain, illness, and mental feebleness. Aside from some peripheral glimpses of spry old men, or matriarchs serenely surrounded by large intergenerational broods, literature offers few portraits of contented old age. Rather, we get the raging Lear, maddened by his loss of power, or the poignancy of Proust's Prince de Guermantes, grown frail and grey at the end of the great opus, or the painful spectacle of the onset and progress of Alzheimer's in Max Frisch's *Man in the Holocene*, or the unadorned, unacceptable facts of illness and death in Philip

Roth's *Everyman*. But altogether, in comparison to other stages of existence, the last stages of life get relatively short shrift in literature, as if the old were narratively, no less than evolutionarily, dispensable, or at least less interesting.

And so, surely, it is to the good that such preconceptions are changing, as we see more 'wonder-elders' leading active and independent lives into their eighties or nineties. Indeed, our perceptions – and the realities – of all periods of life are shifting dramatically from what they were even a generation ago. Maturity, adulthood, middle age and what used to be called old age have all been moved back in time, with a numerical adjustment of at least a decade, if not more. 'Sixty is the new forty' is a heaven-made motto for the baby-boomers, who were never meant to get old and who are reaching their sixth decade just in time to have that former entryway into old age redefined as the height of middle-age powers.

And yet, how far can we defy the imperatives of biological time? Are there bounds to our control over it, beyond which the attempt to bargain with Chronos begins to have diminishing returns? On the sociological and economic levels, the costs of an ageing population have been well reported and rehearsed. An ever-expanding segment of the citizenry living past retirement age creates ever-greater strains on various institutions and social resources: the health services, benefits budgets, pension funds and, eventually, the economy as a whole. And if, in our quest for longevity, we are collectively climbing towards some post-centenarian standard, we have to contemplate the eerie prospect of societies in which the old predominate, and of cultures as a whole subsiding into senescence – a scenario worthy of Jonathan Swift, who, in *Gulliver's Travels*, imagines the Struldbruggs, a group of elders chosen arbitrarily for immortality and threatening to skew their society's balance to such an extent as to make legislation against them necessary. The immortals, in addition to having all the usual vices, acquire some new ones, 'which arose from the dreadful Prospect of never dying. They were not only Opinionative, Peevish, Covetous, Morose, Vain, Talkative, but uncapable of Friendship, and dead to all natural Affection … Envy and impotent desire are their prevailing Passions' – a good premonition, perhaps, of the intergenerational envy which might prevail in a society dominated by the old.

Still, arguments from demographics or the collective good are unlikely to persuade most people that they should deprive themselves individually of the benefits of prolonged life. Perhaps more convincing are the actual miseries which can follow from artificial life extension. We can now keep people in a comatose or unconscious state 'alive' for long years; we can give 'life support' for protracted intervals to people who are in a state of physical disintegration and chronic pain. Many such sufferers have pleaded to be allowed to die with dignity, creating – in a dark dialectic – a pro-euthanasia movement, to counter our culture's death-conquering drive.

This is the Faustian wager turning bitter, or even tragic; the technological defiance of death leading, paradoxically, to a form of death-in-life – a stoppage of meaningful existence through the artificial simulation of vitality. The concrete costs of artificial life maintenance are clear. Less tangibly, but just as importantly, our broader cultural disavowal of mortality, our ideology of agelessness and the belief that we deserve perpetual youth – such attitudes can also eventuate in disturbing, if more elusive, paradoxes. For all that humans have wished to live for ever, for all that our own extinction is, to most, a frightening and unacceptable prospect, for all that, the knowledge of our finitude has been nothing less than the condition of human identity. All living organisms have their temporal fate – but only we know that we have it. We know that our lives have a forward extension in time, and that, however much we might wish to deny it, we are destined to die. This is a fact so enormous and fundamental as to verge on platitude, but much of human endeavour and meaning have followed from it. If we want to make sense of our days, if we want to fill them with something more purposeful than mere existence, if we wrestle with our own significance and insignificance, that is because we are conscious of our own impermanence. Myth, religion and philosophy have arisen from the need to reckon with our awareness of mortality. We have created fables of the world's origins, of the afterlife and of eternity in order to imagine measures of time larger than our own and to counteract the fears of our own ending.

In more secular epochs, works of art and edifices of thought have been prompted by the need to grasp the nature and purposes of our lives in the light of final extinction. Given our impermanence, what possible significance can we attribute to our existence? And yet, one could also ask: what would be its significance if it were truly permanent? What would be the sources and possibilities of meaning for creatures who do not have to face their own end? One of the best cautionary tales about the hazards of immortality is elaborated in Simone de Beauvoir's novel *All Men Are Mortal*. This is an allegorical story of a man who is immortal – and to whom, therefore, nothing, and no one matters. The novel's protagonist cannot take risks, since the ultimate risk of death is eliminated for him, and all consequences of his actions will eventually be cancelled out; he cannot make meaningful choices, for the same reason; he cannot love anyone, since he knows that people are infinitely replaceable. 'I looked indifferently at the evidence of fires and massacres,' he reflects. '*After all, what does it matter*? … The dead were no longer, the living were alive – the world was just as full as ever, the same sun shining in the same sky. There was no one to feel sorry for, nothing to regret.' Others are convinced of the uniqueness of their loves, or the truth of their causes. He is 'a man of nowhere, without a past, a future, without a present. Each instant destroyed the next, the last. I had nothing to hope for.'[10]

Of course, this is a picture of unattainable permanence; and yet, de Beauvoir's intuition of ultimate anomie is suggestive for less extreme conditions – and for what happens if we try to deny our mortality and elude the progress of time. Humans have always nurtured fantasies of overcoming death. No less a realist and a stoic than Sigmund Freud thought that it was impossible to imagine our own individual extinction. But such fantasies have always been known to be fantasies. If, however, we believe that we can exercise temporal omnipotence in the real; if we think we can manipulate time at our will, and reverse its consequences; if we refuse to acknowledge our mortality in some part of ourselves, then we risk – paradoxically – lessening the meaningfulness of our lives. If our experience matters, if we feel that our actions and our selves count, if our choices are accompanied by a sense of consequence, and the people in our lives are precious – that is because we know that each moment is unrepeatable, each stage of life comes around just once, that the losses we incur are real and are to be mourned. We cherish others, and perhaps ourselves, because we know that our lives move in only one direction, and that we are all vulnerable to the workings of time.

Such realisations – such knowledge – can induce fear and trembling, or, at the very least, a deep sense of poignancy. The lexicon of protests against death is vast. In *Nothing to be Frightened Of*, one of the latest additions to the canon, Julian Barnes details examples of ordinary relatives and famous writers receiving what he calls *le reveil mortel*; that is, the full awareness, usually accompanied by full dread, that death applies to them personally.[11] But it is the very knowledge of their mortality which spurs them on to register their rebellion – or resignation. It is the tension between the wish to be immortal and the recognition that we must die from which the effort to understand – to grasp what it is to have a specifically human fate, and to grasp such life as is given to us – proceeds.

Mortality is the prerequisite of meaning, and meditation on mortality may be the best compensation for its fact. 'Birth, copulation and death' was T. S. Eliot's merciless summary of human life. If we are capable of making these brute facts less brute, that is because our consciousness of them is added to their bare factuality, because of our need to fill our fate with significance, and perhaps even with compassion. The passage of time introduces poignancy into each human creature, and sometimes, calls forth in us a corresponding tenderness.

The disavowal of our ultimate fate, on the other hand, requires us to stop grappling with our actual human condition and situation. In order to believe (as, in our technologically enhanced culture, we increasingly do) that we can arrest the movement of time in our veins, that we can reverse ageing and stop death at the pass, we must also, to some extent, deny the possibility of frailty and loss, the consequences of our choices, the precious uniqueness of others' lives, and the continuation of time beyond us, into the next generations.

These may seem like rather subtle reckonings, but eventually the illusion that we can take full control of lived time results – with a sort of ironic inevitability – in a curtailment of what might be called existential vitality, a sort of stoppage of essential human meanings at their source. In any case, the forward movement of time cannot be stopped; that is, perhaps, the best reason for accepting it, and acknowledging the inescapable power of both Chronos and Thanatos. If we do so, we have a chance of negotiating with time on a more realistic basis, and making use of it according to our purposes and to the best of our abilities.

The only way to arrest the passage of time on the biological level is through what has been called the 'complete but reversible cessation of metabolism' – that is through the temporary suspension of all those physiological processes which add up to the experience of being alive. Certain animals resort to strategies of metabolic arrest naturally – for example in hibernation, during which the rate of metabolic processes slows down radically. But the possibility of such biological stop-time has also been demonstrated experimentally by freezing certain organisms for a delimited period of time. Occasionally, humans caught in extreme situations have come close to being frozen, but have come back to consciousness and organic health. In the interval during which the metabolic processes are on hold, the organism does not age; nor do humans experience the movement of time. As Michael Guppy and Peter Hochachka put it, metabolic arrest constitutes 'a reversible escape from time'.[12]

So it is in our existential attitudes to time. The problem with attempted escapes from it is that they occur at the cost of being alive in any recognisable sense of the term. In Joseph Heller's *Catch 22*, one of the novel's protagonists, a soldier caught like the others in a terrible war and terrified of the daily spectacle – and prospect – of death, famously tries to slow down time by making himself very bored. But the joke, as he well knows, is on him, for he makes himself, in effect, bored nearly to death. Time deals us the most basic catch-22s of all, and it cannot be so easily outwitted, either in its forward progress, or in its cyclical demands. In Richard Strauss's tender and sensuous opera *Der Rosenkavalier*, the ageing Marschallin, who knows that her younger lover is destined to leave her, stops all the clocks in the palace so as not to be reminded of the passage of time. The gesture is poignant precisely for its futility. You can stop the clocks, but you cannot stop time. The only way to arrest its progress in the body is to diminish the sense of being alive; the only way to stop it definitively is through that irreversible cessation of metabolism which is death. To be alive is to feel the passage of time, and to have time working through us in every cell, nerve ending and organ, as it takes us through its paces and plays in our bodies its mortal, vital tune.

2

TIME AND THE MIND

Consciousness

It is our consciousness of time, then, that complicates human transactions with that element. Time works through the bodies of all creatures, but only we know that it does so, and that it has an existence – however immaterial – outside ourselves. That is, we perceive time mentally as well as being shaped by it biologically.

'In the beginning ...' In the beginning was the Word, but even before that, there was the assertion of time. Nabokov dates the birth of his consciousness to his first awareness of time, a moment when he felt he was 'plunged abruptly into a radiant and mobile medium that was none other than the pure element of time'. But for all of us, time is to the mind what air is to the lungs: invisible, ubiquitous and absolutely necessary. Without it, our ability to perform mental acts would collapse, as lungs collapse without oxygen. Our first act of orientation in the morning is to ask what time it is – where we are in the day and how we should pitch ourselves towards it. And it is hard to imagine any human act or endeavour that does not depend on the ability to conceive the existence of time beyond the immediate moment. We could not have intentions or decide to go out of the house, or build dwellings or voyage across the sea and savannah, without some projection into the future. We could not recognise ourselves each morning as the person we were yesterday without the mind's constant reach into the past through memory. We could not even register the awareness of a fleeting moment without the perception of time's fleet progress.

The body is subject to the workings of time; the mind recognises time, as well as being constituted by it. Still, what is this ethereal substance, and does it have substance in the first place? What is this thing called time? Does it actually exist and what is it made of? Can we know it in itself, or see it only as through a glass darkly? Such questions are among the oldest and most perennial subjects of philosophic speculation and wonder. It is impossible, and perhaps unnecessary, to give even an abbreviated guide to the philosophical concepts of time, but certainly thinkers from Aristotle to Immanuel Kant and after have grappled with the beast, and have pondered its nature. St Augustine gave the foundational expression to perplexity as he struggled to articulate what the past and the future are, given their non-being – or, at least, non-presence. 'What then is time?' he famously asked. 'If no man ask me the question, I know; but if I pretend to explicate it to anybody, I know it not.'

Ever since then, time has been considered, in philosophic terms, as one of the fundamental categories of experience, or reality; and one particularly concentrated burst of philosophic speculation on its subject came at the end of the nineteenth and the beginning of the twentieth century, from a school of philosophy known as phenomenology. If the propositions of the phenomenologists – Husserl, Bergson, Merleau-Ponty, Heidegger – still have resonance for us today, it is because they tried to analyse perceptions of time as a function of consciousness and subjectivity, rather than an absolute, objective reality. The locus classicus of temporal cognition would have it that time has an external nature, and that it is up to us to recognise and understand it, however limited our minds may be in trying to do so. But in the phenomenological formulations this putatively straightforward relationship between external and internal time was complicated in various ways, and sometimes entirely severed. One of the formative questions of phenomenology was whether our colloquial and intuitive sense that time 'flows' is an objective fact – or an illusion created by some part of our perceptual apparatus. Edmund Husserl, in a seminal formulation, proposed that the sense of flow is achieved through a tripartite structure of temporal cognition. The perception of any immediate moment, Husserl thought, always contains elements of 'retention' and 'protention' – of what has just passed and what is about to occur. The 'now' cannot be prised out of the sequence of the immediate past and the immediate future. For example, we could not hear a melody as melody if our immediate apprehension of the note before our ears was not accompanied by our 'memory' of the note just before and an expectation of the note to follow.

Husserl was trying to pinpoint the way in which our impression of time's continuous flow, of its seemingly unanalysable fluidity, was in the first place constructed by consciousness from pointillistic temporal particles. Henri Bergson took this even further, suggesting that our sense of time was produced not so much from discrete perceptions as through changes in our internal states, and that even the conviction of our own continuity – of a seamless, stable self – was achieved only by melding together different inner occurrences, or motions of subjectivity, through duration and temporal extension. Yet more complicated propositions came from Merleau-Ponty, who suggested that we come into the world equipped with a certain mode of consciousness, so that we are predisposed to understand reality in a certain way. At the same time we are inextricably part of the world which we come into, so that we imagine or experience time and space as we do – *as* time and space – by virtue of being constituted as we are, and of coming into the world as it is. Moreover – hearkening back to a long tradition – Merleau-Ponty stressed that we construct time not through pure perception but through motility and action, through our bodies' motions in space and through acting within and upon the world.

Consciousness, in the phenomenological accounts, was no longer a passive receptacle, receiving the imprint of time; rather, it had its own structuring powers, working on reality, and sometimes interacting with it in a complex dialectic. It is surely not coincidental that phenomenological notions of 'internal relativity' were being put forth at the same period as Einstein's theories of relativity in the external cosmos. It must be said, however, that the philosophers' powers to convey the deep and minute processes of the mind in the impressionistic vocabulary of sensory experience were being strained to the limit. Indeed, and it is tempting to speculate that this is why it was necessary to find a way to speak about consciousness – and the consciousness of time – that was more precise, or at least more suited to the microscopic scale of activity it described.

In the twentieth century explorations of mental temporality shifted to the fields of psychology and psychoanalysis, with their extremely close observations of human subjects; and in our own period, the cognitive aspects of consciousness have become largely the domain of neuroscience, where they are investigated with the help of up-to-date scanning and imaging techniques and through direct probes into the central organ of perception: the brain.

The neurological mapping of temporal consciousness is at its early stages, with the brain on the whole remaining, in many of its regions, terra incognita; but what is being discovered through its explorations echoes, in interesting ways, the more empirical intuitions of the phenomenologists, and indeed of other thinkers who have reflected on this ineffable subject. As in the earlier philosophical systems, time in neuroscience is taken to be a fundamental category, and our consciousness of it a distinctive aspect of human identity. In *Wider Than the Sky: A Revolutionary View of Consciousness*, Gerald M. Edelman, a Nobel-winning neuroscientist, defines higher-order consciousness as 'having the awareness of the past, the future and the self'. [1] Other animals have perceptions of present instants and, in some cases, memories of short durations; but only we, apparently, can envision time's longer vistas, and only we can summon them deliberately to mind – that is decide that we want to recall the past, or project our thoughts into the future.

However, even though time seems to be as insubstantial as ether, and even though the mind can seemingly glide through it resistless, the deep probes into the brain suggest that our perceptions of time are in the first place structured within us, and that, although they have great flexibility, they operate within specific parameters and constraints. Beginning with the instant, or the immediate present: within the brain, even this basic unit of time has to be actively constructed. The brain encodes sensory input from the external world through the firing of neurons, which travel along the synapses to make links with other neurons and eventually create mappings of whole areas of the brain and the distinctive network of connective pathways within it. And it seems that even our most basic and seemingly incontrovertible sense that time is unitary and absolutely smooth in its texture may itself be thus achieved, as part of a wider phenomenon known as integration. This refers to the sensation that any one moment, or scene we perceive, is made of one indivisible piece. What we see is inseparable from what we hear, and both have an absolutely uniform texture. But this conviction of qualitative unity is a kind of *trompe l'oeil*, accomplished through a correlation of various neurological signals – auditory, visual – working together in synchrony. Our basic sense that time moves seamlessly and without demarcations may well be part of this totalising, integrative effect. The brain 'binds' temporality within itself, through its own synchronising processes. Without these, we might experience time as a pointillistic series of events – somewhat like film frames before our vision blends them into continuous motion.

Moreover, it seems that the present is never just the present. The 'now' is as difficult to isolate in itself as a point in the trajectory of a bird in flight. William James, in an effort to catch this effect in a phrase, referred to the 'specious present'. Edelman refers to the 'remembered present' – because, it seems, the perception of each present instant literally depends on the memory of our immediate and accumulated past. Each new impression is added through dynamic 're-entrant' interactions – that is neuronal firings coming from various areas of the brain and connecting various pathways – to the memory networks which the brain already contains. Eventually the new additions become part of the neurological system, modifying or reinforcing the mappings of the brain's neurological networks and informing, in turn, subsequent impressions and signals the brain receives. As in Husserl's tripartite structure of temporal perception, each new 'note' of experience is altered by the context of the preceding sounds; and each will modify the meaning of the notes to come.

This 'bootstrapping between memory and perception',[2] as Edelman calls it, itself takes time – milliseconds to seconds – to be accomplished. This minuscule but nevertheless measurable interval, then, can perhaps be taken to represent the basic temporal unit of consciousness. The production of each thought requires the transmission of signals along neuronal pathways. In order to travel among these networks, the signals need to leap over the gaps between neurons, and this requires the action of certain neurotransmitters. All of this takes actual time. Indeed, Jean-Pierre Changeux, a leading contemporary neuroscientist, has written that 'everyday experience leads us to suppose that thoughts pass through the mind with a rapidity that defies the laws of physics. It comes as a stunning surprise to discover that almost the opposite is true: the brain is slow – very slow – by comparison with the fundamental forces of the physical world.'[3] Although thought seems to travel at the speed of light – or at no speed at all – both thought and perception are, in fact, constrained by the temporality of neurological processes. Beyond a maximum speed of mental operations, we – or the brain – get confused, and do not, for example, register succession. If we are shown two images, or words, one after the other at an interval shorter than the brain can process, we will not be able to tell which comes first and which second. (Perhaps this is the brain's own equivalent of the uncertainty principle, giving rise to the possibility that, at a sufficient acceleration, a micro event can occur both 'before' and 'after'. And perhaps this also accounts for the not uncommon experience of instantaneous premonition – the briefly eerie feeling that we know the glass is about to break before it actually does so.)

As with thought, so with the more externally oriented

processes of attention. If we decide to speed up attentiveness by dividing it among too many signals or demands, the very effectiveness of the brain's functioning is diminished. This is confirmed by a number of experiments which have been performed on the brain's reactions to one very contemporary form of activity – multitasking. It seems that the shift from one activity to another literally takes brain-time – up to seven-tenths of a second. This is because of a 'rule-activation' process whereby the prefrontal cortex activates the set of 'rules' (the appropriate neurological connections and pathways) for each activity. The switch between one activity and another requires a deactivation of one set of rules and the reactivation of another. In one test of multitasking, the subjects underwent PET (positron emission tomography) scans while simultaneously listening to spoken sentences and mentally rotating pairs of three-dimensional figures. Apparently, the scans showed a 29 per cent reduction in brain activity following from this dispersion of focus – so that it took the subjects not less time but more to perform each task.

A similar loss of time and quality of performance has been discovered in experiments when two activities, rather than mental tasks, are alternated, as was found in a study conducted by Peter A. Hancock at the University of Central Florida in Orlando.[4] That means that if we try, say, to cook a complicated new dish, and make an important transaction on the telephone at the same time – as which of us has not? – we are likely to do each less well. And sometimes the consequences are dire. For example there is mounting evidence that texting while doing something else – not only driving but crossing the street, for example – leads to a substantial incidence of accidents. These, for obvious reasons, most often involve young people, and are sometimes fatal.

In order to perform a task properly, or to complete a train of thought, we need to give it a proper amount of attentional energy and focus – which is tantamount to giving it time. This is in the short term. What of the specifically human capacity to conceive longer stretches of time, extending into both the past and the future?

Neuroscience is relatively silent on the latter, perhaps because the future, in terms of neurological signposts no less than in our colloquial conception of it, is more of a blank. It seems that the prefrontal cortex is involved in 'tagging' a sense of expectation, or futurity. Brain imaging is beginning to reveal aspects of perception which indicate the brain's projections into the immediate future – and these seem to involve the faculty of attention. For example some theorists have found that there is a correlation between attention and the ability to estimate upcoming temporal duration. Researchers such as Joseph Glicksohn posit the existence of a 'cognitive timer' in which 'time units' accumulate at various subjective speeds. If our attention is directed outwards, taking in many stimuli, or 'watching the time' actively, the timer is more attuned to external temporality and our estimation of prospective duration is likely to be more accurate. But if our attention is turned inwards, or if we are fully engaged in what we are doing or experiencing, fewer subjective 'time units' accumulate in our consciousness and we are likely to be less aware of duration itself – how much time we have to complete a task, for example, or for how long we have been involved in it. In states of extreme concentration, such as are achieved in meditation, the pulse of subjective time slows down considerably.[5] But, of course, the time 'lost' in such absorption may well be repaid, or regained, in the quality of experience, or what we can achieve within it.

Possibly the capacity to turn our attention in one direction or another implies that there is an active decision, or intention, involved in its deployment – and surely the element of aim contains in itself an element of futurity. (It may also carry in it a sort of synaptic seed of, or a neural opening into, free will.) We choose one target of awareness over another because of what we wish to do, see, or even think about in the next moments. We choose to execute an action or complete a task – and we have some sense of the upcoming time it will require. We imagine that something that has not happened yet is about to happen, because we intend it to happen. We imagine a potential result. 'Each state of consciousness possesses an intentional horizon which refers to potentialities,' Husserl wrote, and some neuro-phenomenologists are picking up on the idea.[6]

The past, by contrast, has once had a palpable existence, and leaves more concrete traces in the brain. The faculty through which the once-existent stretches of time are retained in the mind and brain is, of course, memory, and the neurobiology of this fascinating mental power has been intensively studied in recent decades. In this area, it seems that we share certain capacities with other animals; other forms of memory, however, are particular to us. Many animals may have an equivalent of 'procedural memory' – that is memory available to us non-consciously, and enabling us to perform certain tasks (like riding a bicycle or typing or, for that matter, eating) semi-automatically, without exactly having to 'recall' them. Certainly animals 'remember' how to hunt their prey, for example, and which species are on their menu. Moreover, speculative research suggests that some animals may also have versions of autobiographical or 'episodic' memory, which until recently was thought to be possessed exclusively by humans. This is a form of personal or personalised memory, in which the mind remembers not only certain events or information but also one's own presence at the time of the original occurrence. For example if we remember not only a multiplication table but where we were when we learned it and who taught it to us, we are exercising autobiographical memory. Some recent studies suggest that not only primates – as we would surmise – but scrub jay birds, for example, may have this kind of recall, though the results are still contested.

But we are the only species capable of deliberately deciding to recollect the past; of reflecting on it, revising it and sometimes lying about it. Here we enter a realm of great complexity, and also a kind of innate or inbuilt creativity. For what researchers, with the help of their probing instruments, have been discovering is that the brain is hardly a passive or inert receptacle of impressions which it encodes or stores away, to be retrieved in an identical form later. Rather, the brain – no less than the mind – processes experience with a dynamism which is intrinsically plastic and constructive. The development of our long-term temporality is accomplished through 're-entrant' or variously reciprocal interactions among the brain's pathways and areas – complex cross-signals which inscribe both continuity and change into the underlying structures of our being. The basic spatio-temporal template of the brain is laid down early in our development, but its topography is subsequently modified throughout our lives. Each new experience or piece of information alters the neurological patterns we carry inside ourselves, and the specific encoding of each new impression depends on the context of what is already there. The brain selects, categorises and assigns different 'values' to memories continuously, in the light of its own neurological history. But it is also continuously changed by the new input, or experiences, entering its domain.

Moreover, the re-entrant interactions within the brain enable us to recall our own memories – and to recognise them as our own. The structure known as the hippocampus, located in one of the temporal lobes of the brain, is also implicated in the transformation of short-term perception into long-term episodic memory of sequential events. This makes the hippocampus the guardian of our personal recall – the ability to know that we have had certain experiences, and that we had them in a certain sequence. If the hippocampus is removed, the subject can remember events prior to the removal, but not after. Personal memory, and with it the capacity to know our own temporal continuity, is lost.

That sense of continuity is, of course, crucial to our sense of identity, of that conscious self-awareness which Edelman considers one of the distinguishing hallmarks of human consciousness. What happens when we lose memory's thread has been well recorded in case studies of patients with Alzheimer's or neurological disorders. One vivid account of massive memory loss has been given by Oliver Sacks in his essay 'The Lost Mariner'.[7] Jimmie, the patient described there, probably suffers from a version of Korsakov's syndrome: he has very good recall of events before the accident or trauma which led to his disturbance, but nothing thereafter – including experiences he has just had moments before. His powers of immediate observation, and to some extent the ability to react to what is in front of him, are intact. But he does not recognise the doctor, whom he has seen minutes before; nor can he recognise himself as the same person who had been in the room. In a sense, Jimmie doesn't feel the distress of his condition, for he has no way to be aware of it, but there is nevertheless an objective pathos, or even tragedy, in his situation. Through the loss of his post-traumatic, short-term memory he has become entirely lost to himself, living in discrete moments which have no relation to each other – or to his personhood.

The conviction of time's seamless flow may to some extent be an illusion; but a sense of temporal continuity – the ability to make connections between past and present, and the ongoing stream of experience – is one of the prerequisites of a genuinely human identity. Sacks quotes A. R. Luria, the great Russian neurologist, who describes Korsakov's syndrome as 'Gross disturbances of the organisation of impressions of events and their sequence in time.' As a consequence, patients suffering from Korsakov's 'lost their integral experience of time and begin to live in a world of isolated impressions.' [8]

Clearly, an extreme fragmentation or destructuring of time can become a torment, a nightmare in which we cannot make even minimal sense of our own experience. In order to know ourselves as selves at all – in order to perceive ourselves as subjects and sources of meaning – we need to re-cognise our existence over time.

But in normal circumstances, that sense of continuity – of our own history – is inscribed into the very structures of the brain, and constructed through the micro processes of neuronal circuitry. It seems that each person's specific (and unique) narrative is created in the tracery of neurons and synapses, and the specific connections between them, and that the kinds of modifications of that narrative which we might call interpretation and revision are also built into the material dynamics of the mind. Each person's neuronal network, under the stimulus of experience, develops its own connections and mappings. Eventually, the brain develops a store of its own imagery which it can 'look at' and recombine in new ways, independently of sensory input. This is what happens in sleep for example, and also when consciousness is turned inwards, as we summon certain memories or move mentally over them, reviewing past experiences or viewing them from new perspectives, or perhaps making new connections among past perceptions, which amount to speculations or hypotheses. Such capacities for plastic response and change have led some neuroscientists to speak of the brain's 'autopoiesis' or poetic self-creation. Certainly it is tempting to see in the registration of memories, and their selection, a kind of neurological inventiveness, which, through a combination of stability and plasticity, composes the ongoing narrative or poem of our lives.

Beyond that self-recognition, and the ability to recall our own memories, we humans have the ability to conceive a past which is not constructed from the materials of our own experience, and which precedes our existence. We are able, in other words, to imagine both a historical and an abstract past. Perhaps this happens through the brain's ability to move over its self-contained memory-repertory, leading eventually to more abstract mappings of pathways and networks. But whatever the physiological processes involved, this is surely a feat unachievable by other animals, and one which affects vast areas of our understanding and the nature of our cultural, if not physiological, development and evolution.

Indeed, the conception of time outside the present moment is surely a formative act of abstraction altogether. In the apprehension of a 'past' and 'future' we recognise the existence of something that is not present to our senses as concrete reality. But such detachment from what we can see, smell or hear is the basic condition of thought altogether – at least, of symbolic thought, as opposed to immediate sensory perception. In order to think at all – in order to manipulate symbols independently of sensory data – we have to step out of the full immersion in the physical and temporal surround. The findings of neuroscience suggest that a certain understanding of temporal order is crucial to the development of language, and that concepts of event and succession must exist in a child's mind before the exercise of language can proceed. Indeed, both words and syntactical units unfold in succession; and presumably their unravelling from undifferentiated sound, and their logical ordering, requires at least a minimal distancing of the mind from the encompassing present.

The ability to conceive time, then, may be a crucial constituent of thought, but cognition itself is subject to temporal laws. No matter how much we may feel that our thought takes weightless flight, or that its velocity transcends time, mental processes work within biological materiality and have actual duration. So with the apprehension of time's more extended intervals, or periods. If the 'remembered present' takes some milliseconds to be constructed, the conversion of brief impressions into long-term memory requires more mental work, and longer intervals, as the brain goes through the processes of embedding new information in more permanent pathways or structures. Neurons have to be fired in order for new connections to be made, and certain neurotransmitters have to be produced and conducted through the brain in order for the transformation to occur. Moreover, as anyone involved in 'brain work' will testify, concentrated thought consumes not only time, but energy. It is just possible that, as people have their own metabolic patterns, so individual brains have their own rhythmicity. G. H. Hardy, an English mathematician at the beginning of the twentieth century, specified rather precisely that he could do truly creative work only for four hours a day. The rest of the time he happily played cricket. Einstein apparently needed at least ten hours of sleep a night after his Herculean efforts. A well-known writer has found that she works in ninety-minute segments, and that at the end of each such interval her brain clicks to a stop, as if a switch had been turned off.

Thinking, attention, perception sometimes seem to be so disembodied as not to be subject to any limitations or laws. And yet, if we try to push our minds beyond a certain speed or duration, the circuits get scrambled and basic mental activity fails. In the long run, if we do not give ourselves time to process our experiences, and allow immediate perceptions to settle into memories, the brain – so some researchers speculate – may actually become more dispersed and 'superficial' in its mappings, losing some of its deeper connectivity and experience-shaping powers. In one experiment, researchers at UCLA found that when their subjects were asked to perform two tasks at once – ostensibly, to speed up thought and attention – their brains coped 'by shifting responsibility from the hippocampus, which stores and recalls information', to an area of the brain 'which takes care of rote, repetitive activities'. The subjects involved in this exercise could just about perform their tasks; but they could not remember their content afterwards. The information they had taken in had been more shallowly stored, and quickly evaporated.[9] More generally, in an essay which amounts to a confession of a rehabilitated multitasking addict, Walter Kirn notes that 'the mental balancing acts that [multitasking] requires ... energise regions of the brain that specialise in visual processing and physical coordination and simultaneously appear to short change some of the higher areas related to memory and learning'.[10]

Our conception of time – of the present moment, the past and the future, and the links as well as the discontinuities between them – is initially formed by our own perceptual apparatus, and encoded in the amazing internal machinery which enables and constrains our cognition of the world. The elasticity and swiftness of mental processes is remarkable, but all of them have their neurologically inscribed parameters, and a pace which cannot be accelerated without loss of function or meaning. If we play a record at excessive speed, the sounds emanating from it do not add up to a melody or music. If we hammer a nail into the wall too quickly, we are likely to stub our fingers and put ourselves out of commission.

This is another neat paradox which Chronos serves us: in the mental domain, as in the body, even as we try to get the maximum use out of every minute, we end up getting no use out of them at all. On that level, too, if we do not give time its due, we may short change ourselves by diminishing, or even subverting, our specifically human perceptions, and mental powers.

Psyche

Just as we are not purely biological organisms, so consciousness itself is not pure, or insulated from the body or other aspects of the self. If it were not for that repertory of complex feeling and affective response, buried memories and internal conflicts which we have come to call the psyche, we might be merely – or fully – ratiocinative creatures. But we are not. We have brains, but these are embodied; we have minds, but we also have subjectivities, and it is within the subjectivity that we experience our experience most directly and wholly. It is in our felt sensations that we know ourselves to be not only bodies or minds, but persons; and it is through the dynamics of our interiority that what happens to us matters, that our lives acquire personal valence and significance, that we are not only abstract, but specific selves.

Of course, the very divisions between these parts of the self are to a large extent artificial. Our thoughts would be meaningless if they were entirely impersonal; our physical sensations would be mute and brute if they were not filtered through soul and mind. What, exactly, the connections are between cognitive consciousness and affect, brain and feeling, neuronal pathways and passions, is also, these days, the subject of intense scientific investigation. The well-known neurobiologist Antonio Damasio, together with others, has suggested that emotions arise out of feedback loops between the brain and the body, as our minds scan and react to external impressions, or internal imagery of events; that as our minds check the instinctive or physiological reactions of our bodies to external stimuli (fight, flight, churning stomach, heart flutter) and contextualise them, the more modulated texture of emotion is produced.[11] In another register, cases of cognitive impairment – such as the massive loss of memory described in 'The Lost Mariner' – seem to offer clues to the profound interconnections between cognition and subjectivity. When parts of the brain are damaged, some of our subjectivity, our particular personhood, is diminished or lost.

What is certain is that our perceptions, our apprehension of the basic realities and our dealings with the world, are inescapably inflected by our inner lives – and our transactions with time are no exception. Our temporal awareness may first of all be registered in the brain, and arise from it, but our actual experience of time, our sense of its tempo and duration, have a markedly psychological or emotional aspect. We are nearly all capable of telling clock time, and dividing hours into minutes, and those into seconds. But experiential time is almost never so neatly quantified. It is more peculiar, more laden with sensation, and suffused with valences of pleasure or displeasure. It is accompanied by 'qualia' – the name scientists give to the scientifically still ungraspable qualities and textures of experience. We all recognise the basic axioms of subjective relativity. Time passes quickly when you're having fun. Time spent waiting for crucial news can move with excruciating slowness. Our inner Chronos can be felt as benign or menacing, can trip dancingly from minute to minute or lie heavy on our hands;

it can bring back the past in a flash, or shuttle from the present to a notional future, or meander in other, less easily charted directions.

But in addition to such passing states, most of us exhibit a deeper or more characteristic stance towards time, a set of attitudes and apprehensions which might be called a temporal temperament. Some of us are perpetually aware of time, others seemingly oblivious to its passing. Some live in eager expectation, looking towards time to come as if it held only the promise of good things; others await the future in apocalyptic dread, or blank it out altogether. Some are obsessively punctual, others always, and seemingly helplessly, late. Franz Kafka expressed his temporal temperament perfectly when he wrote, 'Only our concept of Time makes it possible for us to speak of the Day of Judgment by that name; in reality it is a summary court in perpetual session.'[12]

The emotional inflections – the qualia – of inner time suggest that temporality is deeply linked with other aspects of our personalities and selves. But our sense of time (or our temporal sense) does not spring from our heads fully formed. Like all other aspects of selfhood, it requires development and learning. On one level, the necessary development is cognitive. Within the machinery of the brain, the neurological connections need to reach a certain level of stability and complexity for memories to be laid down, and for the perception of extended time to emerge. On the experiential level, the understanding of temporal intervals also grows gradually, as can be seen even in informal observations of young children. A very young child does not know what 'an hour' is. It is immersed in each moment, and may play without being aware that time is passing at all; and it may just as spontaneously abandon its play without knowing how much time has passed. An older child may grasp how long an hour is, and may enquire impatiently when it is going to be up, but it may not yet know how many such intervals there are in a long and tedious afternoon.

Some of the most systematic attempts to study the development of time perception in children were undertaken in the early twentieth century by the cognitive psychologist Jean Piaget. By asking children to perform a series of operations on images which showed flasks emptying and filling with water, and then to order them in a temporal series, Piaget identified several stages of time perception, ranging from the 'sensori-motor' phase, in which the child is entirely immersed in his environment, and is not able to classify a temporal order at all, to the advanced 'operatory stage', in which the child can arrange series of events with reference to an objectively calculable notion of duration – that is, to an abstract idea of 'time in itself'.[13]

Since Piaget did his seminal work, other phases and phenomena have been noted in the development of time perception. Apparently, babies are first aware of time through pulse and rhythm. (Here, too, motion is all.) Adolescents' perceptions of time, perhaps under the influence of hormonal factors, exhibit great volatility, fluctuating between the sense that there is much too much time and that there is no time at all (but perhaps these are only adolescents as we know them). Piaget's work itself has been criticised for being culturally specific, rather than universally descriptive. Not many citizens of the globe, the critics point out, perceive time in such rationalistic or abstracted ways. But even if they do, the ability to measure time coexists with other, more affectively, and perhaps physiologically, coloured perceptions of temporality – ways of shaping time within ourselves that are deeply subjective, and deeply implicated in other dimensions of personal experience.

That inner time and its shaping has been explored most interestingly in the field of psychoanalysis; and it is there that we can find the most fruitful suggestions as to the origins and formation of lived temporal experience. This is not to say that psychoanalysis has habitually addressed the subject of time head on. Mostly, at least until recently, it has not. But questions of time and timelessness permeate psychoanalytic theory, and are embedded deeply within it. Indeed, one scholar of psychoanalysis, Adrian Johnston, has gone so far as to suggest that 'psychoanalysis is, fundamentally, a philosophical insight into the subject's relationships with temporality'.[14]

It is the great contribution of psychoanalysis and related fields of knowledge to show that, while the recognition of objective time is indispensable to us, in the inner world time is constructed from the materials of feeling, and that these rarely follow the laws of chronology, or of measurable, external time. Indeed, it was one of Freud's foundational hypotheses that in some parts of the psyche – the unconscious, or that storehouse of unconscious drives and fantasies, the id – time does not exist at all: 'We are astonished to find in it [the id] an exception to the philosophers' assertion that space and time are necessary forms of our mental acts,' Freud wrote in 'The Structure of the Unconscious'. 'In the id there is nothing corresponding to the idea of time, no recognition of the passage of time, and (a thing which is very remarkable and awaits adequate attention in philosophic thought) no alteration of mental processes by the passage of time.'[15]

The analogies with the supposed state of the universe before the Big Bang are hard to resist; and just as cosmologists make inferences about that unimaginable pre-time from the behaviour of various forces after, so Freud made his hypothesis about the timelessness of the unconscious on the basis of examining his own and his patients' dreams and fantasies, in which, he believed, the contents of the unmodified id are most directly expressed. And both mental processes, he observed, ignore the basic laws and the linear progress of time. A few minutes' dreaming can give the impression of covering great stretches of time. In other instances, dream time becomes so condensed and compressed that the logic of succession seems to be suspended; the present and the past get inverted, or smashed together, as if there had been no intervening passage of time. In fantasies, all our wishes are satisfied without effort or delay. In both, material from long ago surfaces as if it were acutely present, and becomes intermixed with something that happened only the day before. Mortality, in night-time scenarios, is banished, as the dead return to vivid life. And in daydreams, we never age.

Such signs from the unconscious undoubtedly indicate our deepest and dearest wishes: that time should not pass and that we should not die. In effect, they suggest that we do not wish to live *in* time. But in time is exactly where we do live. And, more than any other material in the world, time resists our wishes. Therefore, the child must learn to adjust to it and accommodate itself to the temporal rules which obtain in its environment.

Within psychoanalytic observation and speculation, such adjustment happens not only through the growth of cognition, but through the child's relationships to its intimate others, and to its own impulses and desires. The infant initially exists in an undifferentiated, timeless state; and perhaps it is not unreasonable to conjecture that the earliest moulding of temporality happens through close contact with intimates: through the way the child is held, or rocked, and physically contained; through the tenor and length of a parent's gaze; through the rhythms of its caretakers' gestures and their voices. Rhythmic pulse is the earliest form of time the baby recognises, and by rocking babies, or cooing to them, adults instinctively resort to rhythmic repetition. If the adults' movements are consistently calm, this will undoubtedly communicate itself to the child's still disorganised, receptive mind; if they are habitually hurried or anxious, this will affect the child's own actions, and its still pliable sensations and inner movements.

But in order for more autonomous consciousness of time (and self) to emerge, the child has to develop a sense of its separate being, of containment within its own nascent body and mind. That, in turn, requires at least a rudimentary ability to delay the satisfaction of impulses, and to feel safe within its own frame, in the absence of others. A baby cries when its mother recedes from view, presumably because it does not yet know there is something beyond the moment of absence, that the mother will reappear. If it feels the impulse to eat, it demands through crying that the impulse be satisfied immediately. And, of course, it cannot postpone its bodily movements through self-control.

Gradually, however, the child learns to bear longer intervals of its caretakers' absence, as it begins to know that mother or father will reliably return. It also acquires bodily disciplines which require an adjustment to external temporal rhythms. For example it is taught to eat at certain times, rather than whenever it feels a pang of hunger, and to regulate its bowel movements. The child learns, in other words, that it must accept the temporary frustration of its desires. But it also begins to understand that those desires are usually satisfied after a while – after some time has passed.

In a famous piece of child observation, Freud describes his little grandson repeatedly throwing a toy out of his pram, to the accompaniment of the German word *fort* (gone), then hauling the toy back into the pram by its string, while saying *da* (there!). Freud describes this '*fort-da*' game as the child's attempt to master the loss, or absence of its mother, by the symbolic motions of disappearance and reappearance. But the game is also an example of a child experimenting with time and trying to tame its passage, or bring it under its own control.

It may be through such symbolic negotiations, as well as through neurological development, that the child begins to bear its separateness from others for more than a brief interval – or, conversely, to contain within itself larger intervals of time. And it is through such deeply affective lessons that external time eventually begins to exist as an entity in itself, and that its passage comes to be understood as something that has to be lived through and accepted. For in a sense, nothing happens between desire and its satisfaction – or frustration – except the passage of time. Even if a child is looking forward to a treat, or, when older, a vacation, it still has to contend with the interval before its arrival. Later, if the child is anxious about an exam and wishes it were already over, it still has to bear the time in between.

In the normal course of events a young person will eventually learn to appraise external time more realistically, and to plan activities with regard to its availability. A schoolchild of a certain age, however sulky or resistant, knows that some time in the day must be apportioned for homework. Those of us who live in cultures where punctuality is highly valued also need to learn that if we do not want to be late for an appointment, we have to prepare and set out for it in time. This also involves the understanding that we may have to adjust to other people's schedules, even if this does not suit our immediate desires or impulses; that we may have to keep an appointment or go to school, or show up at the office, even if we 'don't feel like it'. This is a form of discipline, but also (in the time-observant cultures), a basic form of respect for the autonomous existence of others. Recognising that others' time is valuable, that they have their own arrangements and preferences, and that sometimes (or often) we have to accommodate to those preferences, even at the cost of our own, requires a realisation that we live in a shared temporal world, and an adjustment to a reality outside of ourselves.

Realism, among other things – or above all things – is temporal realism, and it seems that in its achievement, no less than in the cognitive consciousness of time, separation and differentiation are crucial moves. It is one of the attainments of maturity to give structure to time: to recognise its parameters, scales and limits. In its full exfoliation, a realistic attitude to time involves not only a grasp of the twenty-four-hour span, but an understanding and acceptance of longer periods and intervals. After we have lived through a number of annual cycles, we incorporate into ourselves a sort of internalised knowledge of what a year is, and how much time it contains. By watching ourselves and others progress through various ages and stages, we acquire a larger sense of the arc and span of human life. In cultures that value long-range planning, some become capable of it, considering how to map out not only days or seasons, but the phases of life, or a career.

In an essay entitled 'Existence in Time: Development

or Catastrophe?' the psychoanalyst David Bell points out that 'feeling oneself as existing in time is an important developmental achievement'.[16] The achievement can be expected if all goes well. But what happens if something goes wrong in the process of psychic development, especially in its crucial early stages? Let us imagine, to give a hypothetical but not implausible example, that the child, for some reason, is unable to bear the intervals of separation from its mother, and feels a terrible anxiety when left on its own. (The mother kept disappearing for too long; or on the contrary, never left the child at all, not allowing it to develop a tolerance for aloneness.) Let us suppose that such anxiety persists later in life, in subliminal and transformed forms, even though the person subject to it may not fully understand its source. Such a person might well develop a distorted sense of time – or evade the awareness of time altogether. We all know people who have great trouble getting out of the house, as if leaving the enclosing nest were tantamount to a frightening ejection from all safety; or who are known for staying at a party till the bitter end, or until they are thrown out – as if they wanted to remain in the warm bath of others' presence for ever. Then there are others who always leave that party prematurely, as if afraid to spend too long in anyone's presence, or of being stifled by others' invasive proximity (which may perhaps summon up earlier, psychically dangerous invasions).

The aetiology of psychological behaviour is, of course, always specific and complex, but the general point is that our orientation towards time is deeply ravelled with our orientation towards others. A person who is insufficiently differentiated from others tends to swim in time as in a watery medium with no demarcations or sequence. At a certain degree of symbiotic merging – if a child never achieved any independence from its first significant others – the grown-up may not achieve any sense of control over time, or the capacity to manage it. Time, in such a mental scheme, is not one's own, and the person who is unconsciously merged with others may be perpetually at their beck and call, shifting plans and schedules to accommodate others' demands. Or else one can imagine a child which was pandered to so fully that it remains arrested in fantasies of omnipotence. Later, this person might find it difficult to be punctual, since this involves meeting someone else's demands and needs. Indeed, the demand to accommodate to external time altogether can, in people with certain psychological profiles, arouse anger or great anxiety.

At the other extreme is the person who has been excessively disciplined as a child – for example in the regulation of its bodily movements, or in being left to cry without consolation or succour. It is quite possible that such a person can maintain a sense of self only through obsessive self-regulation and attempts to impose severe discipline on others. This kind of personality is met with less frequently these days, perhaps as methods of bringing up baby become kinder and gentler, but the character is familiar from literature. It is the paterfamilias of Thomas Mann's *Buddenbrooks*, who demands that dinner be served at precisely the same minute each day; it is the pedant who plods through detail after detail without being able to enter into the subjective content of the material; or the 'control freak' who can never change his plans, or relax into unstructured play, or allow for the play of emotions. For such a person, time is measured in strictly calculated units, as it marches on in a predictable, monotonous rhythm.

There are, in psychoanalytic literature, numerous variations on the theme of neurotic temporal behaviour. To cite just one: Paul Williams, in an account called 'Making Time: Killing Time', talks about a patient who is habitually late for her sessions, and who has 'a fantasy that her mother and I wish to leave her'. The analyst's interpretation is that 'by keeping me waiting and finding me here she experiences the reassuring feeling that I do not intend to leave her. The idea that her mother's life, and mine, are independent of hers is disavowed, as is acknowledgement of her own rejecting feelings.' [17]

In more extreme disorders, the connections between subjectivity and external time can be almost completely disrupted. In clinical depression, time becomes arrested in a hellish stasis, as the sufferer remains convinced that the anguish of this condition will go on for ever; that nothing can ever change. The phenomenology of such states has been well and movingly described in various personal accounts; more statistical approaches show that 'an average depressive, in estimating duration, experiences time as moving twice as slowly as normal.' [18]

In mania, on the other hand, time hurtles along at breakneck speeds, collapsing night and day – people in manic states are known for making middle-of-the-night phonecalls – and accelerating speech until words collide with each other, and sentences and sense crash. At the further reaches of psychotic thought, the fundamental logic of continuity, succession and simultaneity ceases to be recognised, as time becomes compressed in a sort of nuclear fusion. In schizophrenia, a sense of temporal scale is entirely lost as the sufferer enters a self-enclosed temporal cocoon. In one account of a schizophrenic case, the patient thought after several months of treatment that only a day had passed, whereas a day seemed to be only a few minutes. For others, time sometimes seems to go backwards, or literally to lose its flow, as it breaks up into disjointed particles of perception.

This is how the nineteenth-century American poet Emily Dickinson, who may have come close to very radical states of mind herself, described their anguish:

I felt a Cleaving in my Mind
As if my Brain had split –
I tried to match it – Seam by Seam –
But could not make them fit.
The Thought behind, I strove to join
Unto the thought before –
But Sequence ravelled out of Sound
Like Balls – upon a Floor.

Psychic pain puts time out of joint, disturbing its pace and proportions. Indeed, following the classical tripartite structure of time, one could perhaps devise a temporal typology of psychic malaises, dividing them into disorders of the past, present and future.

In psychoanalytic writings, disorders of futurity appear mostly as anxieties about potentiality and change – essentially a kind of phobia of time's forward movement. This can present itself not only as externally oriented worries about 'what the future will bring' – a dread of catastrophic events, say, or ill fortune – but as a recoil from internal development. Psychoanalytic studies abound in cases of people who, in some part of themselves, deny the passage of time because they are afraid, perhaps, of entering into full sexual relationships, or relationships *tout court*, or who do not want to abandon fantasised 'good parents' and the apparent protection such internalised images afford them. In some cases the disavowal of change exists in conflict with other parts of the psyche, as the person moves back and forth between being aware of external time and escaping into 'psychic retreats' – inner spaces where safety inheres in the world being kept out, and in maintaining (at whatever psychic cost) the illusion of timeless stasis. When the denial of external time and change is more complete, rationality cannot be sustained; and the person engaging in such affective strategies begins to suffer delusions, or enter psychotic states.

Potentiality is a form of the unknown, and therefore openness to it requires an acceptance of uncertainty, and of risk – including the risk of loss. But the idea of futurity carries with it one formative piece of knowledge, which is hardly uncertain: the awareness that we, and our loved ones, will die. That inevitable outcome is surely the ultimate hazard, and the ultimate source of future-dread.

Here, however, psychoanalysis complicates matters by positing, in addition to the fear of death, a desire for it. Indeed, the fear, in this formulation, may proceed precisely from the desire. But the wish for death, Freud and subsequent commentators tell us in a further twist, is in a sense also a resistance to change. Such a wish has at its root a fantasy of escape from the conflicts and complexities of ongoing change, and a reversion to a less demanding condition of being – a state of non-consciousness and rest which precedes life, as well as succeeding it. For to live in time is to accept the demands of hope and disappointment, of promise, struggle and loss. The only way to arrest time subjectively is to achieve the psychic equivalent of 'reversible metabolic cessation' – the condition of affective stasis and near-death which is reached in depression and other time-denying states. To flee the hazards of potentiality, of the future tense, is to flee the movement of life itself. And it is through the awareness of loss – and its ever-present possibility – that, psychoanalysts tell us, an awareness of time, and its forward directionality, returns to the disturbed psyche.

It is, however, in describing the relationship of the psyche to the past – that more actually existent aspect of time – that psychoanalysis offers its most fully elaborated insights. In a sense, psychoanalysis started as a theory of memory and its discontents. One of Freud's seminal insights, from which much of his thinking grew, was that memory isn't always what it seems to be; that what we remember overtly about our own past may correspond only partially, or not at all, to what actually happened; and that there are forms of remembering of which we are not overtly cognisant, but which nevertheless exist within us by other means, and affect us powerfully – sometimes all the more powerfully for not being accessible to lucid consciousness. 'Hysterics suffer mainly from reminiscences,' Freud famously opined, by which he meant that hysterical symptoms (paralysis of limbs, inability to speak, etc.) were actually expressions of events, or fantasies, which took place in the past, and which were disturbing enough that the patient could not acknowledge them explicitly. Symptoms were, in a sense, enactments of repressed memories – vital clues to a seemingly forgotten past.

Repression causes a kind of blockage in the normal flow of time, and its recollection. Psychoanalytic literature offers a compendium of terms for the strategies of disrupting internal movement between past and present – and vice versa: fixation, repetition compulsion, regression. All of these are ways in which the psyche becomes arrested at a painful point in time, or responds to the past's excessive pressure. A person who has undergone a traumatic experience in the early stages of development may remain fixated, so to speak, upon the moment of trauma, and upon the states which the experience produced: anxiety, excessive rage, excessive sexualisation. Such a person may repeatedly revert to the memories of the terrible event, or their unconscious representations, as if no time has intervened, and nothing has happened in between. In the more developed syndromes of repetition compulsion, temporality becomes stuck like a snagged record needle, going over the same automatically reiterated notes. It has been noted that in compulsive-obsessive disorders, the sufferer cannot imagine forms of movement *other* than repetition. Or, under the pressure of internal conflicts, a patient may revert to earlier, more childish forms of response and feeling, almost literally reversing the direction of time by actively moving into an earlier self.

Sometimes the fixation on a painful past happens as a result not so much of neurosis as of extreme circumstances. This is often the case in the aftermath of catastrophe, or of adult trauma. Childhood trauma freezes time by blocking access to the past. But it is increasingly recognised that trauma which occurs in later life can also fix time at the moment of terror. In studies of Holocaust survivors, for example, this motif recurs again and again: for many of those who endured horror, time, for all intents and purposes, stops at the point of greatest danger or loss. Many are haunted by nightmares in which the frightening episodes surface with a piercing vividness; for others, the present is blanched of all meaning, as the past keeps returning with an overwhelming presentness.

The future can be an object of dread; the past can weigh heavily on the psyche, or press on it with a phantasmal power. But can there be such a thing as disorders of the present? In a certain vein of thought, 'living in the present' is seen as a great desideratum, a way of asserting freedom and the right to pleasure. But more often, the contraction of time to the present signals psychic danger or extremity. Dostoevsky, who was placed in front of a firing squad, ostensibly to be executed, experienced such concentration on the immediate moment that it seemed to be filled with all life's richness. He resolved from then on to relish each 'now' as if it were sufficient in itself, but without the prod of extreme danger he could not maintain this pitch of intensity for long. For inmates in concentration camps, the curtailment of time to the present was a form of temporal, in addition to physical, incarceration. The past, for them, was cruelly severed, with no hope of return; the future was foreshortened by the looming wall of probable death. The exit from temporal confinement was blocked in both directions, making 'living in the present' a form not of liberation, but of terrible psychic imprisonment.

But in recent decades, observers have begun to notice, and to decode, disorders of the present which are of a more systemic kind, and whose symptoms reflect not anomalous extremity but certain pervasive cultural predispositions and values. This is particularly true of a syndrome referred to as borderline personality disorder (BPD) – a malaise which, in its present version (rather than rarely met antecedents), started to be widely diagnosed around the 1960s and which has since become so prevalent as to be an almost normative psychological – or cultural – style. The borderline personality needed to be identified and described in the first place as a result of psychoanalysts encountering patients whose internal organisation no longer corresponded to the classical categories of id, ego and superego, and whose subjectivity never cohered into stable or continuous internal structures. Instead, the subjectivity of such patients remains radically disorganised and fragmented, and, without any psychic basis for self-control, they are unable to impose restraint on their impulses, or perspective on their experiences. Their dealings with time are similarly dispersed. Characteristically, such people drift or skip from one event or encounter to another, seeking ever new sensations or gratifications without being able to give them context, or endow them with any non-immediate significance. It is typical of BPD patients that often they cannot recall what they have done, or why what they did mattered to them. Nor can they make connections between different points in their temporal trajectory. In the description of Thomas Fuchs, a psychoanalytic theorist, they 'exhibit a fragmentation of narrative self', engaging in a 'temporal splitting of the self that excludes past and future as dimensions of object constancy, bonding, commitment, responsibility and guilt'. [19] It can be said that the past in such subjectivities is not so much distorted as never constructed in the first place.

The torment of this very modern condition comes not from such relatively outmoded feelings as, say, guilt about one's past actions, or anxiety about not being able to meet the standards of one's superego. Rather, the baffling pain of BPD is that disparate experiences remain entirely dissociated from each other; that sensations never accumulate into personally felt meanings; that ongoing relationships cannot be formed or maintained; that a sense of identity remains perpetually provisional and tenuous; and that even self-recognition from day to day and experience to experience is uncertain, in a psychic equivalent of Korsakov's syndrome. As Fuchs writes, 'The loss of time as a continuum that extends into the past and the future creates a now without depth' [20] – and without personal meaning.

This is neurotic suffering incurred in the service of narcissistic gratification. The paradox is cruel, but the catalogue of BPD symptoms adds up to a very familiar character type, and many commentators have noted how closely the syndrome's manifestations parallel the wider disorders of late-capitalist cultures, with their segmentation of experience, loss of shared ethics and emphasis on immediate satisfaction of impulse and desire. (*Shopping and Fucking*, a 1990s play by Mark Ravenhill, may have delivered one of the definitive titles of our era. And it is notable that people diagnosed with BPD often resort to alcohol and drugs, so easily available and so easily promising instant satisfaction.) Fuchs straightforwardly attributes the rise of BPD to the fragmentation of modern life and the breakdown of stable bonds in our 'pluralistic, mobile, nomadic' cultures – prefiguring, or perhaps paralleling, the somewhat later advent of attention deficit disorder, with its even more extreme fracturing of perception and temporal sense.

In other words, BPD is seen as a disorder of the times, as well as of time. But the description of BPD as a malaise of temporality – as much as of drives or desires – is in itself relatively new, and interesting in its implications. Does the prioritising of time as a diagnostic category offer clues to other psychic pathologies? Usually an unravelling of temporality has been seen as a symptom of more structural psychic changes. But are temporal anomalies themselves causes of mental disturbance? There are at least some observers who are beginning to hazard the idea that this is so – that if our sense of time is sufficiently awry, we can go good and mad as a result. For example certain chemical and hormonal imbalances which affect our biological clocks have also been noted in depression and anxiety. The neuropsychologist Kim Dawson sees temporal disorganisation as a breakdown in 'the integrative relationships between the various (neurological) representations of time', and she points out that such breakdowns can be caused by chemical agents such as drugs, or by normal processes such as ageing. This suggests that the causality of mental disturbance does not have to be purely psychic. As Dawson puts it, 'The wide range of mental illnesses associated with temporal breakdown strengthens the likelihood that time is an important aetiological factor underlying mental illness.'[21] There may well be looping feedback mechanisms – as there are in so many aspects of our inner functioning – between the construction of time and the structures of subjectivity. Loss or mourning or early trauma may lead to attempts to evade time or hold it still; but conversely, if the fabric of temporality is itself sufficiently torn, in either individual chemistry or the culture at large, a shattering of psychic structures may be the consequence. Pain puts time out of joint, but if time is sufficiently out of joint in the first place, great pain may follow.

~

In its more moderate versions, the predilection towards one of the three dimensions of time – past, future, or present – may express itself not as neurosis but as a sort of temperamental predisposition, or existential style. There is the traditionalist who cherishes the past and wants to study it or preserve it, or return to it in nostalgic longing; there is the future-oriented personality, which envisions optimistic scenarios for personal improvement, or creates models of ideal societies; there is the perpetual modernist, who values all that is novel and experimental; and there is the lyrical sensibility, which seeks moments of heightened intensity, Wordsworthian 'spots of time'.

However, the excesses of pathology are often a clue to the normal, and what the more extravagant disorders of temporality suggest is that inner time too may have its unwritten laws, or norms, and that these are very different from the rules and regulations of clock time. In the normal psyche, as in the mind under stress, inner temporality rarely follows the logic of chronology, or a linear, unidirectional arrow. Rather, even as we adjust to the demands of social and external time in our actions, inner time moves within more multilayered and multidirectional topologies – folding and unfolding from the moment to full extension, winding and unwinding from fast to slow, conflating past and present into one perception, or uncoiling into the distant future with gliding ease.

Felt time comes accompanied by qualia; both pleasure

and pain travel along temporal currents. The French psychoanalyst Jacques Lacan speaks (à la Husserl) about the constant movement of the psyche between 'retroaction' and 'anticipation'[22] – but these, in psychic reality, are always charged by the energy of feeling and desire. Small intervals of time are inflected by the gap between wish and fulfilment, or measured by the backward gaze of regret or self-appraisal. If we feel a confident anticipation that our expectations will be met, we may experience the intervening time as equable and spacious; if we fear what will happen next, the minutes or hours in between may constrict with anxiety. (It seems that in the inner world, perhaps by some analogy with the space–time continuum of the Einsteinian cosmos, the modalities of space and time are at least metaphorically inseparable from each other.)

Each moment within ourselves is sensually and affectively constructed. But if we are not to be creatures of the moment, driven – or riven – by impulses we don't understand, we also need to make sense of the psychic flux, and to make links between our felt past and lived present; and this cannot be accomplished without the more overarching processes of introspection, self-examination and reflection. Above all, the construction of the more extended time within us happens through the unpredictable, often digressive, sometimes involuntary processes of affective memory. (It was the taste of a small cake which suddenly brought back to Proust the sensuously detailed expanses of the past, and gave him his great opus. No wonder he was a great believer in involuntary memory.) As both neurologists and psychoanalysts know, memory, when it is not blocked by physiological damage or unbearable psychic pressure, is a highly plastic interpretative faculty. Even in its more neutral cognitive function, memory – on the empirical, as well as neurological level – is necessarily selective. We forget the majority of impressions which have registered on our consciousness; if we didn't, if every errant sight or sound remained lodged in our minds, we could not make much sense of our ongoing experience – a point made rather terrifyingly in Borges's 'Funes the Memorious', the story of a man who remembered absolutely everything. In the ordinary course of events we forget facts or names which are not especially important to us; apparently such data is pushed into some nether region of the brain, and eventually erased in order to make room, closer to the surface of consciousness, for new impressions and perceptions.

But for all of us there are more emotionally coloured forms of remembering – and forgetting. In the short term, memory plucks out of the stream of time particular pebbles of experience which have some special significance to us. We may delete from our minds the time we treated someone badly, or an important anniversary about which we have very mixed feelings. Other events may be recollected with excessive frequency, in a kind of over-remembering. The mind may keep throwing up a moment when we suffered a social embarrassment or an annoying slight in niggling detail ('It was the scornful way she looked at me … how could she?'); or it may return again and again to occasions of unusual excitement or personal triumph (the iterative recollection of each episode in a new love affair is a symptom well known to the sufferer's friends – as well as to the sufferers themselves).

In its long-term operations, too, the relationship of memory to actual events is complex, shifting dramatically or subtly as we ourselves change. Again, psychoanalysis offers a compendium of concepts. One of Freud's more famous terms, *Nachtraglichkeit* – usually translated as 'deferred action' – refers to the delayed action of early drives and instincts on our mental lives and, conversely, the delayed action of understanding on early experience. It is only through retrospection, Freud suggests, that our most formative, early experiences acquire meaning, but this is to a large extent true of later experiences as well. We live forward, but we understand backwards. And, as we acquire new experiences, or new perspectives on the old ones, as we sometimes expand our understanding or deepen our insights, so the interpretation of the past can change over time. The entire picture of our childhood may fall into a different configuration, as we 'reframe' our view of it. The mother who seemed overly disciplinarian, say, may become an object of greater sympathy when we come to understand the causes of her behaviour; the sense of grievance towards a friend who rejected us may lighten when we understand our own provocative behaviour in the light of greater maturity. This is a sign, in a sense, that the psyche isn't stuck, and the ability to return to the past in order to apprehend it more richly, or in proportion, can be an experience of great satisfaction – the development, from within, of a new kind of knowledge.

Memory constructs the past, and reconstructs it. This is not a question of revising the facts but, as with the modifications in neurological pathways, of placing experiential information in new – sometimes larger and richer – contexts. In that sense, the past changes under the pressure of the present, as well as vice versa. The narrative composed by our subjectivity rarely either follows a chronological sequence or is divided into regular chronometric units. Rather, emotional memory, like its neurological correlates, entails acts of continuous autopoiesis, or creative self-making, and the compositions it creates over time, and with the materials of time, are plastic, personal, our own.

It may also be said that the reach of subjectivity – as of cognition – sometimes extends itself beyond the boundaries of our own lifetime, through the kind of 'remembering' which happens across generations and individual psyches. Cross-generational memory has been much studied lately, mostly in relation to catastrophic events (particularly the Holocaust), and the impact of traumatic parental experience on children. Repeatedly it has been found that, even if the children themselves did not experience the originating events, they nevertheless absorb their emotional sequelae in potent and palpable ways. In a sense, they are formed by those events. The children may not initially understand what the parents lived through, but they can sense the states of sadness, anger, or loss and mourning which the parents willingly or unconsciously convey. Later the children can learn about the actualities of their parents' history, a past which, for them, has deeply felt and still living meanings. (I always 'knew' about the attic in which my parents had hidden during the Holocaust, or at least had an image of it in my mind; much later, I went on to study the broader picture and sequence of events in which their hiding took place.)

Undoubtedly such cross-generational transmission takes place in less troubling contexts as well, although its impact may be less sharply manifest. The passage of internalised knowledge can take the form of literal family stories, or reminiscences of earlier times. But it also happens in less deliberate ways, through the very formation of parental sensibility, in which children can intuit something of the preceding generation's values or attitudes, of the elders' own internalised, embodied memories. Nabokov writes beautifully about hearing his mother reminisce about her youth and, by sensing what her memories meant to her, adding to his own store of the things he cherished.

In another register, the fables and fairy tales told to children – in cultures where such stories are still told – convey to the child something about the broader temporal topography of the world in which it finds itself. Indeed, it is striking how many children's stories start (like the Bible) with a temporal indicator: 'Once upon a time ...', 'Once, long ago ...'. Such narrations encourage the child's imagination to stretch beyond the confines of the immediately real; to intuit expanses of time much vaster than the individual, biological lifetime.

In such ways, periods previous to our own lifetime can leave their imprint on us and deeply affect our internal constructions of time; and in such ways we can touch on history, not as an abstraction, but as inhabited, human past. It is, moreover, another psychoanalytic insight that, within the inner world, all scales and layers of time can exist simultaneously. Freud frequently used archaeological metaphors for psychic structure, and the psychoanalytic excavations of its hidden strata. André Green, a prominent contemporary psychoanalyst, speaks about temporal 'heterochrony' – the coexistence at any one point of various kinds of temporal strategies: repetition, screen memories, regression, anticipation.[23] We may be, affectively speaking, several ages at once. It is perhaps this trans-temporal aspect of the psyche – the timeless quality of dreams, the gliding ability of fantasy and memory to move back and forth in time – that gives rise, and a sort of dreamlike, uncanny plausibility, to fables of time travel or eternal youth.

The psyche constructs time continuously, and it does so with great malleability. And yet, within a living person – as opposed to pages of science fiction – there are limits to temporal flexibility. What the norms may be in such a fluctuating realm is undoubtedly difficult to discern or define, but the more extreme disturbances of psychic temporality offer clues as to their implicit existence. The excessive expansion of time, or its excessive constriction, are literally hurtful conditions. Emily Dickinson, the anatomist of inner spaces, again:

Pain – expands the Time –
Ages coil within
The minute Circumference
Of a single Brain –
Pain contracts – the Time –
Occupied with Shot
Gamuts of Eternities
Are as they were not –

If time in the psyche loses its scale and form, we court falling into inner chaos. But possibly, between the extremes of stasis and agitated turmoil, there is, for each of us, a proper sense of proportion and pace for subjective processes, as there is for walking or breathing; a right rhythm and scale of lived experience – of being-in-the-world – which we need to find for ourselves for the sake of our well-being, and of being well.

~

But outside the internal relativity of the psyche – and beyond any subjective experience we may have of it – time passes. No matter how we fold or pattern it inside us, how we manage or elude it, or try to stop it, or expand it, it moves on, at its own terrestrial and sidereal pace. No matter what our access to infantile feelings or fantasies, our bodies age. However we may want to speed up the interval between the present and an event a week from now which we are awaiting with trembling anxiety or hope, the week has to pass before the event can occur. In that sense, Beckett's allegories of existence as nothing but time passing – his *Endgame* and *Waiting for Godot* – may be quintessential, if very bleak, representations of time. And, however intense our wish to revert to an earlier event, we cannot turn back in actual time; and we cannot move through it more than once.

That, undoubtedly, is the psyche's trouble with real time: it is we who are ephemeral within its stream; and to acknowledge its movement is, ultimately, to acknowledge our own death. And yet, there is no alternative, not only because of the demands of truth, but because to turn away from the facts of flux and change is to risk psychic paralysis or delusion. The attempts to arrest time so as not to acknowledge loss, to evade it so as to avoid change, or to chop it up so as to avoid its flow, ultimately lead not to pleasure or self-preservation, but to the sufferings of psychosis or depression, or the loss of identity attendant on borderline personality disorders.

There is always poignancy – even sadness – in the awareness of time's passage. This is how another poet – Philip Larkin – expressed it:

> Truly, though our element is time,
> We are not suited to the long perspectives
> Open at each instant of our lives.
> They link us to our losses; worse,
> They show us what we have as it once was,
> Blindingly undiminished, just as though
> By acting differently, we could have kept it so.

And yet, in the realm of subjectivity, as much as of the body, the denial of time can only be achieved at the cost of partial death. In that sense, trying to achieve temporal omnipotence is a losing game. In Greek mythology, Chronos is the first of all the gods, creating order out of chaos but capable also of eating his own children. It is to cope with what time takes away from us, and what it devours, that humankind's first philosophies – its religions and myths – created alternative temporal topographies of eternity and the afterlife.

Freud, the philosopher of the secular age, thought not only that our ephemeral nature has to be accepted, but that it is a guarantee of human meaning. During a brief walk which has entered literary history, Freud met Rainer Maria Rilke – a poet who experienced a terror of mortality and who disconsolately felt that the transience of all things human meant that, ultimately, they had no value; they didn't count. Not so, responded Freud. It is the transience of nature and human beings – of the loved human face – that gives them their poignant significance; it is because we know all things living shall pass that we cherish them. As the psychoanalyst Adam Phillips puts it in *Darwin's Worms*, 'transience is scarcity value in time'.[24]

It is the psychoanalytic intuition that the assent to 'real' time happens in part through valuing others, and an acknowledgement of their full and autonomous existence. The psychoanalyst Paul Williams notes that 'to accept the passage of time means being aware of and concerned for others … To make time is, ultimately, an act of love.'[25]

But perhaps in order to attain this difficult awareness

– to accept our living-in-time – we need first of all to take account of the psychic mode of temporality. In extreme neurosis, we may have to devote quite a bit of time to unblocking chunks of frozen time, and allowing it to flow again. But in ordinary life, if we are not to succumb to illness, or fall into the rigidity of thoughtless routine, we need the space (so often equivalent to time) to make sense of what is going on within. We need to acknowledge the mute motions of our interiority, and catch their drift through reflection, or a sort of inner interpretation. Sometimes we need to pause in order to listen to the inchoate movements of our thoughts and feelings, to let them meander in aimless free association, or crystallise into an unexpected insight. And, on a more extended scale, we need to make links between emotional cause and consequence, and to orient ourselves within the inner topography of our life's events and stages. We need to give time to inner time.

It is in the modernist fictions of Virginia Woolf, or Proust, or James Joyce – works written during the heroic era of exploration of the internal worlds – that we can find the fullest representations of, and reflections on, subjective temporality. That, indeed, is part of these imaginative works' interest and appeal. Psychological narratives mould time in ways that, unlike science fiction, make sense of our perceptions and feelings. Such fictions can give us all the modalities of perception, and valences of time (the fleeting moment, the meanderings of thought, the unveiling of 'time itself' in the middle section of Woolf's *To the Lighthouse*, and the pity of time's passing) in sensuous detail, and with their colours and qualia.

But the fictions are based on materials of perception in the first place; and in our ongoing lives we all, in a sense, need to engage in such acts of creativity. We carry time within us, and we make human time out of our own dark and light materials. And in order to mould neutral time into personal meaning we need to reflect on the contents of our experience, and to filter them through our own sensibility. It is through such processes of self-knowledge and self-feeling that we can arrive at what Paul Ricoeur and other psycho-philosophers have come to call a 'narrative identity' – that is, a sense of self which is derived not only from a purely chronological continuity, but from a significant shaping of our own lived story.

And perhaps, for the full achievement of a personal narrative, we need a kind of perspectival mapping of our own lived time, a balanced vision of its internal topography. If we do not want the past to overwhelm us, or fall into a permanent melancholia, then we need to grapple with its meanings, and absorb it into our self-understanding. And if we do not want the future to transfix us with its menace, then we need a sufficient trust in the world – sufficient psychic safety – to allow that openness to potentiality which is called hope. The inner gesture of 'looking forward' to something is also a psychic achievement.

It may also be in the psyche as in works of art: it is form and structure which give meaning to detail. It is when we can apprehend all the dimensions of inner time, and travel through its strata and striations with some ease, that we can refind, and perhaps reclaim, lost time, and that we can begin to cherish each transient moment for its full value.

3

TIME AND CULTURE

Human time, then, is in important ways subjective. But it is also, just as importantly, relational and intersubjective. Like language, time is one of the fundamental dimensions of human reality; and like language, it mediates between interior experience and the external world. We construct a lived sense of time within, but we are also constructed by it and by the shared temporal order in which we live. In the initial stages, a child's sense of time develops through its relations with intimate others – adults who already embody within themselves certain patterns of temporality. Those patterns, in turn, reflect and are largely created by culture – that system of visible customs and invisible assumptions, unwritten codes and subterranean values which structures, even if we are not overtly aware of it, our perceptions and views of the world. In Western cultures, for example, it is an unwritten but widely understood rule that we need to learn how to show up for an appointment at a mutually agreed time, or to arrive at work at a collectively designated hour.

But beyond such specific codes, each culture has its 'deep structures' of time, which configure the topographies of past, present and future and which extend to ideas of origin or visions of eschatological endings, transactions with ancestors, and rhythms of daily activity.

The variety of such configurations seems to be nearly

as great as the diversity of human languages; and often, to the outsider, difference at this level is as baffling as an unknown speech. In some South American tribes, people point ahead of themselves when referring to the past, reversing our sense that the past is somewhere behind us. In others, there seem to be no repetitive units within the calendar. Rather, the beginning of each season is discerned by tribal sages, or else time is divided into units of days, with each day manifesting itself in particular qualities and properties.

In the more travelled world, contrasts between approaches to time – how much it is valued, how people move within it, how important punctuality is – are the stuff of tourist tales and writerly ruminations. Here is Ryszard Kapuścinski, the renowned Polish journalist and travel writer, discoursing on 'African time' in his book *The Shadow of the Sun*:

> The European and the African have an entirely different concept of time. In the European worldview, time exists outside man, exists objectively, and has measurable and linear characteristics … Africans apprehend time differently. For them, it is a much looser concept, more open, elastic, subjective. It is man who influences time, its shape, course and rhythm (man acting, of course, with the consent of gods and ancestors) … Time appears as a result of our actions, and vanishes when we neglect or ignore it. It is something that springs to life under our influence, but falls into a state of hibernation, even nonexistence, if we do not direct our energy toward it.[1]

He goes on to describe the African capacity for an

'inanimate' waiting, for falling into a state he describes as 'a kind of profound, physiological sleep', which is the result of this temporal sense.

Kapuścinski attributes such attitudes to the African climate, and the ubiquitous, energy-defeating presence of the sun. Whether his diagnosis or his descriptions are entirely accurate or not, every tourist has observed the differences in lived time which coexist contemporaneously within our world – most evidently, between rural and urban, between economically deprived and economically advanced parts of it. Every traveller through the less industrialised parts of the globe has been in rooms or offices where petitioners wait while nothing happens, except, perhaps, the buzz of a fly or the whirring of a ceiling fan; every traveller has witnessed forms of patience with which those in speedy societies have grown impatient.

In *A Geography of Time: The Temporal Misadventures of a Social Psychologist*, the social psychologist Robert Levine recounts a number of sometimes exasperating and sometimes diverting encounters with time-laxness prevailing outside the developed industrial world: university students who show up for classes whenever they feel like it in Brazil, trains which arrive a day late in India and, in one instance of extreme slowdown, waiting three days for a long-distance phone connection in Nepal. But Levine also makes a systematic attempt to chart the differences in pace of life and attitudes to time usage in various cities and countries. In the course of this, he makes some fascinating observations, and finds sometimes unexpected correlations between the pace of life and its quality. By some measures, the United States comes rather low on the tempo index, in comparison to several European and Asian countries (where people apparently move faster, perform tasks more quickly, and have more accurate clocks). At the same time, the citizens of most European countries, despite their quick and agile ways, have more time for what Levine calls *la dolce vita*, and are less hurried and stressed than most denizens of the US. There are also internal differences among cities. Not surprisingly, fastest-paced cities (whatever the indexes of speed), produce not only more stress but less civic generosity. Slower time allows people to be kinder to each other, to take a few moments to help that old lady across the street, or to listen to a shopkeeper who wants to chat about something that happened that morning. (Among the Kelantese people of the Malay Peninsula, 'Haste is considered a breach of ethics,' and part of the cultural code is to make time to visit relatives and friends.) [2] But the relationships between pace and general well-being are sometimes more surprising. It seems that citizens of some 'fast' countries are happier than citizens of some 'slow' ones, and that sometimes economic development is a good index of general contentment.

Within comparative anthropology, cultures are often classified not through tempo but through more structural temporal categories. One often used distinction is between societies and tribes which function on 'event time' – that is, time measured by natural signs, and by how long things actually take to happen – and those

---

which are organised around clock time. (A gathering of tribal elders – as opposed to a corporate pow-wow – presumably takes as long as it takes.) Another central distinction made by anthropologists (and historians) is between two broadly differing conceptions of time's motion: the linear and the cyclical. The 'linear' vision stresses the directional movement of time and the singularity of each event as it occurs, one unrepeatable time, in the temporal continuum. Such an idea of 'time's arrow' encourages notions of development, and sometimes of progress. The 'cyclical' cultures parse time through the seasons, and their repeated return; or, in more sophisticated versions, through the periodic recurrence of certain human tendencies and genres of events (as in Nietszche's idea of 'eternal return'). In the cyclical vision of temporality, events are not so much unique manifestations of singular historical moments as archetypes of larger historical or mythic patterns. In traditional Jewish history, for example, the expulsion from Spain at the end of the fifteenth century was seen as a paradigmatic repetition of the exodus from Egypt. The linear vision may encourage individual effort and will; the cyclical, a belief in destiny, or the Wheel of Fortune, with its gnomes of bad and good luck. It may lead to resignation in the face of events, or an acceptance of time's passage, and the eventualities or accidents it brings.

But such broad categories do not really give us full access to the deeper experience of each culture's lived time. How is the flow of time actually sensed, how is its pace regulated, how are slowness or speed perceived, and how do they reflect a larger vision, or philosophy of time? One fascinating attempt to do a close reading of another culture's temporal map was made by the anthropologist Clifford Geertz in his classic essay, 'Person, Time and Conduct in Bali'. In that extended analysis, Geertz describes a highly elaborated set of cultural conventions, rituals and conceptions which are essentially designed to render time immovable; and to portray the world as existing in an eternal present. This central vision of time is underpinned, for example, by a system of naming in which persons are identified not as specific individuals existent in their particular time, but by their place in a family's birth-order (first, second, and third child), and their location within generations (grandfather, grandchild, etc.). This, and other naming systems within the Balinese symbolic order, depict almost everyone – 'even the dead and the unborn – as stereotyped contemporaries, abstract and anonymous fellowmen'. [3] The use of abstract identifications, within which categories of names (and, implicitly, persons), repeat themselves through the generations, de-emphasises the passage of time in favour of an always-existing order, and creates 'a detemporalizing ... conception of time'. Similarly, the Balinese calendrical systems 'are largely used not to measure the elapse of time, nor yet to accent the uniqueness and irrecoverability of the passing moment, but to mark and classify the qualitative modalities in terms of which time manifests itself in human experience'. [4] The Balinese use a lunar–solar calendar, but also a 'permutational' one, which consists of ten different cycles, varying in length, of day names (recurring ten-, nine-, eight-, and even one-day phases). These cycles run contemporaneously, and important occasions such as holidays – of which there are many – are determined by the convergence of important cycles (five, six and seven). All this intricate complexity serves not so much to 'measure the rate at which time passes', Geertz notes, but rather 'is adapted ... for distinguishing and classifying discrete, self-subsistent particles of time – "days". The cycles and supercycles ... do not accumulate, they do not build, and they are not consumed. They don't tell you what time it is; they tell you what kind of time it is.' [5] There are good and bad days, apparently, 'on which to build a house, launch a business enterprise ... sharpen cock spurs ... hold a puppet show', etc. In addition, each temple – of which there are also many – has its own day of celebration, or 'odalan' – not a birthday, but a day of 'emergence' or 'appearance' on which the gods come down from the heavens to inhabit it.

This is cultural difference at its most radical. The Balinese sense of time, as reflected in its calendrical and generational calculations, is so intricate, and so remote from ours, that most of us can hardly grasp it, or penetrate its lived textures. But in each culture, the temporal order is so deeply bound up with the wider matrix of values, with the conception of the human and its place in the cosmos, as to be tantamount to an existential topography. For the Balinese, a sense of spacious stasis is clearly foundational, and infiltrates every aspect of life in ways which seem very opaque to an outsider. And conversely, it may be that some of our own temporal assumptions, which to us seem so self-evident as not to need stating, may be culturally configured. The very concept of having 'a life' – that is, an individual biography, whose trajectory over time we follow and worry about – may not be universal, and may require a certain idea of separate individuality, as well as a certain level of self-esteem, to make its appearance. The psychosociologist Jerome Bruner, who was interested in varieties of 'self-narratives' – the ways we perceive and recount our lives – suggested that in certain traditional cultures, it is the group's history, rather than any individual 'story' that matters; and that each life is understood to reflect that broader narrative, or pattern.[6] To people in such cohesive cultures, it might seem strange to attribute significance to some odd occurrence in their own lives, never mind to their particular feelings or ambivalences about it.

The perennial question – as with language – is how much difference such differences make, how much they really affect people's behaviour, choices, or social relations. There is a school of sociological anthropology, derived from Emile Durkheim, which contends that the ostensible diversity in temporal cognition merely masks underlying constants; that when all is said and done, everybody has rational notions of real, forward-moving time; and that, as with linguistic syntax, the human grammar of time is structured around certain universals: notions of mortality and immortality, succession and simultaneity, duration and periodicity, among others.

Undoubtedly, we could not recognise each other

across cultural divides as fundamentally human without such a basic grammar of perception. And yet the specific inflections of such universals seem to matter profoundly, both for inward experience and for collective life. As with language, the induction into a specific cultural temporality probably happens very early, through a psychophysical moulding of the child's self, and is therefore deeply encoded. How abruptly or smoothly the child's caretakers handle its body, the rhythm and pace of their speech, or, for that matter, what kinds of lullabies they sing, may begin to shape the child's deep sense of inner time, composing its body and psyche into a rhythm and pace that will eventually seem inseparable from selfhood.

A recent discovery in the field of neuroscience may be suggestive for thinking about the transmission of deep cultural patterns. The discovery has to do with the identification of 'mirror neurons' – that is neurons that fire not only when we perform certain actions, but when these same actions are performed by someone else in our proximity. If someone smiles at us broadly, an equivalent of this gesture is registered in our neuronal firings. If someone near us is breathing agitatedly, we are likely to start doing so, or at least reflect those rhythms internally. In a sense this seems to revive the old idea that we learn by imitation, although the mimicry here is so deep and intrinsic as to amount to a kind of neurological empathy. Some leading neuroscientists believe that mirror neurons are implicated in the learning of language, and surely this mode of transmission may extend to the structuring of time as well. Bodily rhythms – especially expressive rhythms – are something we pick up on most readily, and in which coordination with others is most easily achieved, as shown, for example, in our seemingly natural ability to dance or sing together. And rhythmic forms of expression – in music and dance, as well as in poetry and gesture – are great carriers of temporality; it is within these that specific cultural valences of time are often most directly and intelligibly articulated.

If time is shaped individually within the subjectivity, then its cultural constructions can be said to provide temporal ego-ideals – the normative conceptions, frameworks and rhythmicities which structure our perceptions, or to which we aspire. But, of course, cultures – especially in our fast-changing epoch – change. Indeed, change itself – its pace, modalities and effects on various societies – has in the last decades become a major subject for anthropological enquiry. One suggestive account of cultural temporality in transition is provided in a seminal essay by Pierre Bourdieu, 'The Attitude of the Algerian Peasant Toward Time', written in the 1960s. Throughout much of his study, Bourdieu describes a peasant culture characterised by an almost idyllic harmony with natural rhythms, a philosophical acceptance of time's passage, and an ability to yield to the flow of experience:

> The profound feelings of dependence and solidarity toward that nature whose vagaries and rigours he suffers … foster in the Kabyle peasant an attitude of submission and of nonchalant indifference to the passage of time which no one dreams of mastering, using up, or saving … All the acts of life are free from the limitations of

> the timetable, even sleep, even work, which ignores all obsession with productivity and yields. Haste is seen as a lack of decorum, combined with diabolical ambition … A whole art of passing time, or better, of taking one's time, has been developed here.'[7]

But as the essay progresses, Bourdieu goes on to connect this seemingly bucolic state of affairs with the Kabyle peasants' static social circumstances, and their entire lack of control over them. Once that changes, everything else changes too – as Bourdieu illustrates by citing an experiment in which factory workers from the Kabyle culture were given a dramatic increase in their hitherto minimal wages. 'As if a threshold had been broken,' he writes, 'the workers showed a desire to work, to earn even more, to work overtime, to anticipate the future by thrift.'[8]

What Bourdieu is describing, on a microcosmic level, is a transition from event time to clock time; from a culture of fatalism to one of control. And, in his Marxist or materialist interpretation, the transition reveals that the forces which really matter in the shaping of time – as in all else – are economic and power relations. For the Kabyle peasants, the sense of futurity, the ability to plan ahead and exploit time productively, arises from a change in their economic circumstances which introduces the possibility of improvement, and which therefore makes it worthwhile for them to make a concerted, and perhaps not entirely natural, effort.

It is perhaps possible to say that in advanced, complex and constantly changing societies, economic structures and development have to some extent replaced the older categories of culture and custom as the forces shaping behaviour, and individual subjectivity. This is breeding its own complexes of adjustments and reactions. In Bourdieu's account, the shift from fatalism to futurism is seen as entirely positive. Still, one wonders if the Kabyle peasants hired by the local factory eventually came to feel that, in the course of gaining new advantages, something was sacrificed as well. From the gathering body of evidence it seems clear that an abrupt shift from traditional to advanced regimes of Chronos can be felt as disruptive, and even oppressive. I will not easily forget an Indian businessman I met on a plane, who was travelling from Bombay to New York on what was for him a regular commute. His wife had a fabulous job in New York; he had started up a successful enterprise in Bombay. But he wondered, rather sadly, if all this fabulousness was worth the depletion of his energy, and the loss of marital intimacy and stability. In less glamorous circumstances, the newly hired cadres in Indian call centres apparently find the ruthless temporal regimen to which they are subjected – working often at night, and under strict regulations for the allotment of talk-time – perplexing, as well as exhausting.

This is not to idealise the cultures of fatalism, or to say that those who have made the difficult transition from traditional to modern or post-modern cultures would rather go back to their previous way of life. Very few would. The Bombay businessman wished he could give up his literally high-flying life, but he didn't think he would have the nerve to do it. The employees interviewed in Indian call-centres complained of stress and fatigue, but they almost unanimously agreed that their jobs gave them opportunities they would not want to give up.

Modernity has its compelling attractions as well as its high costs. It can be invigorating, as well as economically desirable, to step into more efficiently organised forms of time. Yet cultural constructions are deeply embedded in the psyche, and there is perhaps nothing deeper than the constructions of time. Being prised out of one's most inward assumptions and ego-ideals can be disruptive in the extreme (witness, for example, evidence of increased risk of psychiatric disorders among Afro-Caribbean immigrants in Britain, or Surinamese and Dutch Antillean immigrants to the Netherlands). The dialectic between various cultural temporalities can produce fertile tensions; but in its sharp and abrupt versions, it can also lead to more violent reactions, and collisions.

4
TIME IN OUR TIME

The question then is, what kind of social, or cultural time do we live in nowadays, and how does this affect the shape of personal experience? What has become of time in our time?

Modern time has its own history and well-charted stages of development. There was the huge shift, beginning in the fourteenth century but probably not fully consolidated until the seventeenth, from telling time by natural signs or events to measuring it by the clock. This had enormous consequences, of course, for the regularisation of all human activity, and especially of work. From the early, not very accurate mechanical timepieces to Greenwich Mean Time, timekeeping has been characterised by ever increasing precision and standardisation; and at every stage, even as clocks with their wonderful and often beautifully executed mechanisms exerted a great fascination, the regularisation of time they represented was resisted as a violation of more natural rhythms. For people who adjusted the pace of their activities to dawn and dusk, the regimentation of the clock seemed a great tyranny. The introduction of time zones, which meant that clocks in towns up to an hour apart longitudinally would show the same time, rather than the 'true' time of the locality, in some places induced vigorous protests. The demands of punctuality, and the incursion of clock time into every sphere of life, were bewailed. People felt squeezed and cramped by the imposition of grid time onto event time; thrown off their own rhythms and equilibrium; unnaturally hurried and harried.

But such standardisation, of course, underpinned all the developments we consider to be part of modernity. Without it, railway timetables would have been impossible to establish, the industrial revolution, with its need for transport and timed work hours, could not have happened. In turn, the creation of industrial technologies, combined with the application of precise time measurements, had two enormous effects: the introduction of new levels and possibilities of speed into human activity and the accelerating loss of personal time. At the height of the industrial revolution labourers in its satanic mills worked up to sixteen hours a day, often in horrific conditions and under the relentless prodding of factory managers and overseers. This was truly the subjugation of humans through the control of their lived time, and it reminds us that having some say over the use of one's own time is one of the most basic aspects of freedom.

By the beginning of the twentieth century, the devotion to efficiency and productivity seemed to induce – and to be dependent on – a sort of measurement mania. Frederick Taylor, 'the father of scientific management', conducted time-and-motion studies in which the physical movements involved in performing various tasks (picking up an object, wrapping one's hand around it) were measured down to fractions of a second. From these calculations, factory managers derived the 'standard time' required for the completion of relevant work at hand, dividing workers' movements into 'waste' motions (such as sneezing or scratching their heads), and those which were necessary for the performance of a task. The assembly line, with its strictly timed production of interchangeable parts and the coordination of uniform tasks and physical movements, was pioneered in this period at Henry Ford's car-making factories. So were production timetables, which required individuals to fit into an overall plan 'like a cog' into a machine (to summon a once frequently used simile). Management came to mean, above all, time management; and while the segmentation and routinisation of tasks enabled ever larger scales of mass production, it involved the control no longer only of the workers' lived time, but of their bodily rhythms and pace.

The prevalence of fast technologies – trains, cars, telephones and telegrams – which could move at much greater speeds than the human body, the standardisation of lived time and its increasing pressurisation, all made for a great transformation in the experience of time. This was perhaps the first temporal paradigm shift of modernity, and it led to a great preoccupation with two new cultural demons: speed and stress. Machine speeds sometimes induced literal fear, as when early trains caused some people to faint or have heart attacks. But speed had its acolytes as well as detractors – for example the Futurists, a group of Italian artists who exulted in the power of the machine and the velocities it enabled as a glamorous, specifically modern form of energy. Stress, of course, was a more unequivocally negative phenomenon, and it led to a spate of books on 'modern nervousness', describing the ill effects of time-pressure and hurry and the ways to counteract these in daily life.

These preoccupations are still with us, but the ante has been upped to such an extent as to make for another paradigmatic shift – a qualitative transformation, not only in the levels of rush and stress but in the very character and materiality of lived time.

The causes of the transformation are in part economic and sociological. For one thing, in our 24/7 societies, there is no longer any time out, at least in the public space. When I first arrived in London in the early 1990s, the city basically closed down on Sundays and the extended Christmas holiday produced an interval of almost eerie urban quiet. No more. Now in London, as in New York and other world metropolises, the urban beat goes on late into the night and every day of the year, producing constant movement and the possibility of round-the-clock activity. In London, human traffic in the shopping areas reaches gridlock levels even on Sundays. In New York, exercise clubs exhibit, through their plate-glass windows, well-muscled bodies running on treadmills long after the streets outside have darkened.

We can shop till we drop and we can entertain ourselves till the small hours, but there is another, more surprising reason for the escalating pressurisation of time: in most advanced, or advancing, countries, people work harder – or at least work longer hours – than they did throughout much of the twentieth century. This may seem counterintuitive; but there are convincing statistics to show that, for average Americans and Britons, the number of leisure hours has actually decreased over the last century. In Britain, for example, the number of work hours kept decreasing throughout the twentieth century until the 1980s, and then began to climb again. Partly this has to do with the large-scale entry of women into the workforce. That means that there is less total time, statistically distributed, to devote to 'leisure' proper, but also fewer total hours to devote to household maintenance, child rearing and even relationships – to what used to be called 'private life'. In reality, the brunt of such life sustenance is still carried on by women, so that many find themselves, in effect, fitting two jobs into every day. The same is true for single-person households, whose numbers have risen sharply in the last few decades, although even in those the pressures are especially severe for single women with children. But as men start contributing to housework and child rearing, they are also beginning to lose the benefits which accrued to them from the old division of labour. The father who comes home after a lengthy day at the office and sinks into an easy chair, with a newspaper and a cocktail served to him obligingly by his wife, is a rare spectacle these days.

In any case, for today's professional parent that long day at the office is likely to be even longer. This is in part because white-collar professional work has become ever more competitive, leading to a simple calculation: those who work less hard (or for fewer hours) get left behind. Moreover, in a sort of entrapping irony, the more successful you become, the more it pays to put in those hours. This is because of so-called 'opportunity costs', whereby time spent in not working costs more in terms of money lost. Whereas a minimum-wage earner in the UK or America may forfeit only a few pounds or dollars per hour of unemployment, a successful lawyer may be 'throwing away' several hundred.

This doesn't necessarily mean that those who stay in the office till mid-evening get more done. Comparative statistics show that productivity per hour is higher in some European countries than in the American workplace, even if the total yield for the country remains proportionally lower, because of longer vacations or shorter working days. But in the post-modern workplace, the competition is in part precisely temporal: to see who can put in the most time. After all, having an overfilled schedule is a sign of being in demand; it shows that you have necessary and important things to do. It is the best educated and the most successful professionals who work the longest hours. Doing so is itself a sign of success.

Social norms, once established, can exercise enormous influence on individual behaviour, and in the ethos of the professional workplace, pressure has become sexy. It is a sign of energy, of upward mobility, of worth. Power breakfasts are not for wimps and neither are weekend working hours.

But there is another, more surprising reason for the increase in middle-class working time, at least in the US: it seems that Americans *like* to work. Indeed, they prefer it to doing most other things. Even as they report in large numbers that they are suffering extremely high levels of stress, when given an option of working less they choose to reject it in equally large numbers. Perhaps this simply confirms the notion that Americans are congenital workaholics; but aside from the incentives of upward mobility and pay there is another, newer factor influencing their decision, and that is that life outside the office – at home and at supposed leisure – is at least as much work as work itself.

Some of these findings first emerged in a revealing and much discussed book, *The Time Bind*, by the American sociologist Arlie Russell Hochschild, and have been echoed since then by findings in Britain. In the late 1990s, Hochschild conducted in-depth studies at a large American corporation, Amerco – one of the enlightened companies which professes a commitment to 'work–life balance', and offers its employees flexitime, part-time work, job-sharing and other reduced-work options. What she found was that most employees refused to take these up. On the contrary, a sizeable proportion of them wanted to work longer hours.[1] Such statistics have been replicated on a larger scale as well. In the late 1990s, 88 per cent of Fortune 500 manufacturing firms offered part-time options to their employees; only 3 to 5 per cent made use of them.

The reasons given for such unexpected choices are in part purely competitive – fear of co-workers' resentment, of falling behind and falling out of the company culture. But when Hochschild talked to Amerco's employees and observed their lives, she found that home, for them, no longer meant respite, or a chance to relax. Rather, many saw it as a sort of 'third shift,' just as busy as the other two, with numerous, segmented tasks to be accomplished, kids' schedules to be coordinated in carefully calibrated sequence (piano lessons, exercises, visits to the psychiatrist), and e-mails to be answered late into the night. Home, she found, far from being the traditional place of refuge, is acquiring a 'Taylorized feel', with 'family time ... succumbing to a cult of efficiency previously associated with the workplace'.[2] 'Quality time,' Hochschild notes, is now another category of activity, to be slotted into the schedule and dispatched with dutiful diligence. It's a form of investment in a rubric called 'family life', or 'intimacy', and a kind of commodity, albeit one that is particularly scarce.

America is undoubtedly in the vanguard of the time revolution, but Hochschild's findings have resonance in many parts of the advanced – and advancing – world. In Japan (as in Korea) the work hours surpass those of Western countries, although cross-cultural observers such as Robert Levine, caution that the actual experience of this should not be translated into Western terms of pressure and stress. The more collective and less competitive approach to work in Japan, and the wraparound environment of its corporations, apparently provides a sense of almost familial security and safety. On the other hand, Japanese has a word, *karoshi*, meaning 'death from overwork', a combination which suggests (at least to a remote observer) that people, in their loyalty to the firms which employ them, work themselves serenely to death.

On more familiar ground, Leon Kreitzman in *The 24 Hour Society*, a study of time patterns in Britain published in 1999, finds that 'A large proportion of the British population believe that they are overworked, and that life is out of control.'[3] Few, however, choose to, or can afford to, work less. Rather, as Peter Cochrane, then head of research at British Telecom pithily notes, the contemporary work conditions have created a new class divide within society: between 'those who spend a lot of time trying to save money', and 'those who spend a lot of money trying to save time'.[4]

Whether time or money is nowadays the more valuable currency is one of the questions asked in *Discretionary Time: A New Measure of Freedom*, a study of time usage conducted by several researchers in six advanced countries, including the United States, Australia, France, Germany, Sweden and Finland.[5] It is the study's premise that having some control over the deployment of one's time is a basic dimension of individual freedom, and therefore a basic good, and that 'temporal autonomy' consists in having truly discretionary time – that is, not only hours spent outside of work, but intervals that are not given over to tasks considered necessary and obligatory. By those criteria, the authors argue, the majority of citizens in advanced countries suffer from real shortages of genuinely free time – time which is entirely at our disposal, to manage as we will. We may no longer be 'wage slaves', selling all our time to employers in order to survive, but great portions of our days are spent on activities which are either truly indispensable or which we consider necessary to maintain a respectable standard of life. (In the authors' catalogue, these include sleep and personal maintenance, transport to and from work, as well as the hours given over to the sustenance of household and family.)

The study proposes policies and agendas which could alleviate this situation, such as better child care, or making it possible for people to choose which hours they want to work without penalising them for this flexibility with lower wages. 'Cultures of equality' apparently give their members more discretionary time by minimising wage differentials between the lowest and the highest, thus diminishing the need to compete and the urgency of reaching higher salary rungs. But, the authors also note, the temporal constraints under which we live are partly of our own making. We could decide to do fewer things, or declare fewer activities necessary, or expect less perfectionism of ourselves. This, however, seems to be a choice that most people – at least those signed up to the benefits of middle-class prosperity and upward mobility – are unable or unwilling to make. Moreover, although there are differences in work hours and stress among various societies, the authors find these basic trends in all the countries they examine – even those committed less to high capitalism and more to the small pleasures of life. The French may be forced, by the imperatives of global competition, to work longer hours and give up some of their vacation time. And informal reports have it that even the Spanish siesta is going by the wayside, to the confusion and dismay of its traditional practitioners.

Indeed, it is one of the signs or symptoms of our times that the dichotomies between work and leisure (or, for that matter, work and life) are disappearing, and that we are willing to accept levels of pressure and haste that would have astonished and alarmed those overstressed early moderns. In post-industrial societies the temporal modalities which used to be associated with paid labour – efficiency, effort, strict adherence to clock time – have been transplanted from such labour to all areas of life. For the professional post-moderns, there seems to be no downtime, and no time out.

But, in contrast to the earlier versions of the work ethic, this is not necessarily tied to any set of values that would give our activities purpose and orientation. The Protestant ethos driving industrial development was grounded in the ideology of progress and a linear conception of time. It carried traces of the religious belief with which it was originally linked, and an eschatological vision of time in which earthly temporality inevitably moved towards a future where all our efforts would be judged. The work ethic at its height required great discipline of personality, and the sacrifice of present pleasures for future goals. It involved a systematic commitment to saving money, so that capital would accumulate, as proof of effort and virtue. The capitalist culture was a culture of the future par excellence; it encouraged its adherents to move through time with a long-term goal in their mind's eye – or, conversely, with a sense of severe guilt and even sin if they failed to meet their objectives.

Not so in our own, thoroughly disillusioned epoch. After decades of expansion and its spoilages we no longer find ideas of human perfectibility, or even progress, sustainable. Rather, we seem to be driven by being driven. This is no longer the work ethic but the ethos of conspicuous exertion, and under its aegis we willingly submit ourselves to temporal regimes that would have seemed rigid or even tyrannical by the standards of most other places and periods.

The pressures on lived time in contemporary societies are partly produced by our social arrangements and the thrust of our cultural aspirations. But there is another, more elemental force which infiltrates and alters time's very fabric and form: the creative and destructive demiurge of technology.

We all know the story: mobile phone, video, computers, DVD, the world wide web. In less than two decades the exponential growth in the quantity and capacity of personal communication and information devices has created a new environmental surround, an omnipresent stratosphere or zone, of digital pulses and signals, superimposed upon, or interposed into, our physical environment. At each moment of the day and night we can plug into any number of gadgets and possible sources of information, entertainment and communication. If we don't turn on the TV, we might watch a video or a DVD; if we're on the road, we might call or text any number of people on a mobile phone. We might play some music on our iPod, or peruse e-mails on our BlackBerry. Moreover, while packing multiple quanta of information into each moment, digital technologies further segment temporal units (and disperse our attention) by inserting into them different kinds of data simultaneously. On news programmes, divided screens and ribbons of written information compete for our attention. In the meantime, the patter of talk emitted from television and radio comes at us ever faster and is chopped up into ever shorter sound bites.

The overcrowding of each moment extends not only to our actual activities, but to the activities available to us, all the time and at any time. Each temporal point holds a proliferation of possibilities and claims on our attention, which can be summoned – or dismissed – by pressing the relevant key on the appropriate keypad.

Indeed, not only do we have multiple options for filling each moment, but none of these options need ever be lost. If we miss the film, we can rent the DVD; if we miss the TV programme, we can access it on YouTube; if we miss a significant political article in a newspaper, we can find it on the web. The quanta of information are not only available simultaneously; they are, increasingly, available in perpetuity.

To compound the sense of limitless proliferation, not only do we have a simultaneous multiplicity of information always within our reach, but the simultaneity now extends to the entire globe. Any one minute now contains all the time zones within its own present. We can communicate across time divides in any number of media, make stock-market transactions in the middle of a New York night into Hong Kong's midday, receive news of coups and earthquakes from across the world in real time. We need never be cut off from any of the globe's minutes. My own most vivid example of this came when I called a journalist friend on her mobile phone. I thought I was calling her in Hammersmith – an admittedly remote part of London. But when she picked up the phone I heard turbulent sounds of a roaring crowd in the background. She was, in fact, in Peshawar in Pakistan, and somewhere near her violent political riots were taking place at that very moment. I had stepped briefly into a micro-slice of someone else's history, and just as quickly, the slice receded – as far as I was concerned – into cyberspace.

The deterritorialisation of time is complete. Not only can time now be prised out of nature's diurnal and seasonal cycles as we travel across time zones within our bodies, it can also be entirely severed from the geographical place and time in which our bodies are physically located. With the exception of certain rural areas in less developed countries, we all now exist in several time zones at once, and in the virtual time of no place at all.

In that sense, the perception of time shortages is very real. We know we cannot absorb everything that global or even local time has to offer, or respond to every interesting possibility within our reach, or 'keep up' and 'catch up' with the journalistic, aesthetic and even personal information bombarding us through multiple media. The Polish poet and Nobel laureate Czeslaw Milosz thought the 'phenomenon of number' – by which he meant quantity – was one of the truly new and profound problems of contemporary life. And surely the awareness of the virtual activities occupying each moment together with the ones in which we are actually engaged makes each of those moments seem too crowded, too insufficient, too brief.

Then there is life with computers, in front of which the denizens of the middle classes spend increasing portions of their days. For many of those below middle age it is becoming difficult to imagine life before the PC. Indeed, few of us at this point could, or would want to, do without this magic machine. But the personal computer has introduced new patterns of mental functioning, and new levels of speed, which are altering both our inner and outer worlds in ways we have yet to grasp, or fully understand.

The speeds at which computers process information within their binary innards are measured in nanoseconds and are, of course, beyond any that an individual mind is capable of. This, so far, holds only for information which can be parsed into abstract units, and not yet for sensory input. A computer cannot react as quickly as a human to save a child about to fall off a high chair – although perhaps that capacity will eventually be developed as well.

The velocity of digital processing is abstract, and invisible to us. But it is in its interactions with us – or vice versa – that the computer has introduced an actual speed of function hitherto impossible for comparable activities.

Basically, what we expect from the computer – in writing, in accessing information, in communication and the delivery of various entertainment products – is instant response. If we want a certain item of information, we expect to access it immediately. If we want to listen to a piece of music, we no longer need even to get up and search for the record, tape or CD; we can find it within seconds by pressing a key on a computer keyboard or mobile phone. If we want to alter something we have written, this can be done with an ease and speed entirely impossible with any earlier writing or printing technology.

At the same time, the internet has expanded the universe of instantly available information to nearly infinite size. The universal library of Borges's imagination – containing all sets of ideas that have ever been written or thought, and then containing the set which includes itself – is fast becoming a surreal reality. Moreover, we do not need to travel any distance at all to access it. It is with us, in our homes, as well as simultaneously everywhere and nowhere.

And, of course, if we want to shop on the internet, the world is our oyster. Commercial transactions effected by this means still sometimes take minutes rather than seconds to complete, but the purchase and receipt of physical goods no longer requires the movement of our bodies across physical distances. They too can be achieved by entering a few items of information on the keypad. Moreover, our purchasing opportunities expand to include the globe, rather than the locality where we physically find ourselves.

The computer is shortening the span of both our patience and our attention. It is accustoming us to speeds of reaction and response measured not in hours or minutes but in brief seconds. How long we expect to wait for information to be delivered, how much time we are willing to give to any mental task before moving on, what pace and density of stimulus we need in order to feel that something 'interesting' is happening: all those expectations are crucially affected by the tempo and procedures of fast technologies. (In my early coexistence with the computer I caught myself making impatient finger-snapping gestures when it failed to respond to a command for something like fifteen seconds. I knew then that human time had changed.) We do not like to wait more than a few moments – and these are getting briefer all the time – for our PC or iPod or BlackBerry to perform their operations. New software is being developed to shorten that 'very long' start-up time for computers. Apparently, people can waste three minutes each morning waiting for their PC to boot up – and this is clearly unacceptable. We expect instant gratification.

At the same time, the multiplication of available activity is encouraging states of extreme cognitive dispersal. If attention is partly a matter of intention, a deliberate decision to focus on a mental or physical object – then the sheer quantity of input crowding our mental field makes that basic act more difficult to accomplish. With so many possible stimuli 'fighting for our attention' in each unit of time, it becomes more difficult to choose any one object of interest for exclusive concentration, or to sustain any one mental act over a prolonged interval. Some observers believe that habitual coexistence with digital time is altering not only our mental habits, but the very structure of our brains. The 'new brain' of some observers' speculations is potentially quicker and more capable of multitasking, but it may encode information more superficially and retain it less lastingly than the brain we have so laboriously and evolutionarily developed.

The effects of technology are undoubtedly particularly potent for children, whose brains and selves are still receptive to being moulded by external influences. For the generations growing up after the introduction of digital devices, the contemporary technological surround must seem to constitute a kind of second nature. Children take to computers like fish to water, using them to solve problems, to socialise and, above all, to play various kinds of – usually very fast – games. And there seems to be no doubt that computers have a great educational potential, giving young people easy access to unprecedented diversity of information and inducting them into certain kinds of mental manipulation. But the question is whether the flexibility and agility developed through such operations are acquired at the cost of other forms of mental activity, such as thoughtfulness, imagination, meaningful absorption of – and in – new knowledge. The very modus operandi of the computer encourages 'surfing' on the surface of informational fields; however much fun navigating the mouse and the keypad may be for a playful child, it rarely leads into deep waters. *Homo zappiens* is how the Dutch educationalist Wim

Veen dubs the new generation of technology-savvy kids, and the sobriquet is often used as a term of approval. But it may not be clear for a while what relationship *zappiens* has to *sapiens*.[6]

The structure of temporality created by our interactions with contemporary technologies has led one cultural commentator, John Tomlinson, to coin a new concept: the condition of immediacy. In *The Culture of Speed*, Tomlinson distinguishes 'immediacy' from the 'mechanical velocity' of the preceding era and notes that, within the parameters of mechanical speed, 'the gap between here and there, now and later, what we desire and we can expect to receive, was *preserved* in the necessity of effort, in the application of will and … the prudential deployment of planning and regulation'. Within the reign of current technologies, '*The gap is already closed*'.[7]

~

Patterns of pathology tend to expose the fault lines in each period's ideas of the normal. Given the nature of our temporal structures and surround, it makes perfect sense that the latest addition to the panoply of recognised psychological disorders involves, essentially, an intolerance of extended or continuous duration. Attention deficit disorder, or ADD (sometimes also dubbed, for good measure, as attention deficit hyperactive disorder, or ADHD), has its dim antecedents in late-nineteenth-century psychiatric history, but it first began to be identified on a wide scale less than three decades ago, initially in children and young people – although the diagnosis is increasingly being extended to adults. But then, since the initial eruption of the ADD epidemic, the generations infected by it have had a chance to grow into maturity. The syndrome is characterised by a list of familiar symptoms: restlessness, fidgety behaviour, impulsiveness, inability to inhibit one's immediate urges and a constant quest after new stimulus or sensation.

What such a list of symptoms amounts to is an extreme shortage of patience and an extreme dispersal of attention. The sufferers afflicted by this very up-to-date malady live in radically reduced temporality, unable to maintain concentration or to direct their attention consistently to any one task or activity beyond the immediate moment. The syndrome has been described as a 'difficulty of sustaining a behaviour or thought over a delay'.[8] Brain imaging has shown that people with ADHD have smaller right-sided pre-frontal cortex regions, associated with attention and inhibition.

The symptoms characteristic of ADD no longer pertain to the loss of emotional continuity between one experience and another, as in borderline personality disorder; rather, they suggest a more severe splintering of time and attention, in which disparate impressions do not collect into experiences in the first place. The causes of ADHD are usually described in physiological or genetic terms. Studies have found that ADHD runs in families. Investigations of the chemical workings of the ailment attribute it to insufficient production of dopamine – a key neurotransmitter implicated in states of subjective well-being as well as internal organisation of movement and alertness.

But this does not account for the great and sudden increase in the prevalence of ADHD – a statistical leap which surely cannot be explained by genes or biology alone. Why were there so many fewer children with this syndrome before, say, the 1980s? Why was the malaise first diagnosed on a large scale in America – the advance guard of fast time?

It is, of course, just possible that ADHD in young children is a new name, and explanation, for modes of behaviour which were always with us; that the children thus diagnosed would have been seen earlier as difficult or restless, or even unusually lively, rather than in some way disturbed. It is also possible that with enough parental and pedagogical patience – commodities in short supply these days – such children would have 'come round' and become more disciplined in the ordinary course of events.

But even if we accept the statistical rise in the numbers of such children – and the data suggests that we should – there is a lot of evidence pointing to non-genetic explanations of the illness, enough to allow for the hypothesis that the chemical and physiological deviations found in young people with ADHD are not causative but themselves symptomatic of previous conditions and mental states. For one thing, it has become an accepted fact in 'the biology of mind' that long-term anxiety can change gene expression – in other words, that states of mind can affect underlying physiology.[9] The symptoms of ADHD in children are treated routinely with drugs such as Ritalin (methylphenidate), which help release dopamine. But quantities of dopamine in the brain seem to be strongly susceptible to modification, not only by chemical intervention but by more substantive influences. For example in both animal and human studies the levels of this neurotransmitter go up when subjects are given certain rewards. And in both cases, if the rewards are administered after a task is successfully completed, the organism will start emitting more dopamine in anticipation of the next reward. Eventually, humans who perform well at their chosen tasks release more dopamine even without the proverbial candy at the end; the satisfaction of doing something well becomes (in a rather old-fashioned conclusion) its own reward. Conversely, it is possible that ADHD children emit less dopamine *because* they feel anxious and internally disorganised, rather than the other way round. If they managed to calm down sufficiently to sustain attention and become engaged in various activities, their condition might subside without the interjection of chemical substances.

A rather more quirky but suggestive finding uncovered in some neurobiological studies is that religious upbringing makes young people less susceptible to impulsive behaviour. [10] Although such evidence has not been investigated extensively, one can perhaps conjecture the reasons for this correlation: religion, on the most basic level, provides a strong framework for the organisation of self and inner time. It offers a hierarchy of significance which can allow children and young people to select among the quanta of stimuli bombarding them indiscriminately in our overcrowded cultural space; and it adumbrates a structure of values, symbols and rituals which can be internalised and held within the self, lessening the constant need for external stimulation. In all these ways, religion can be taken as a template of a containing meaning-system. But there are, of course, other possibilities. One can conjecture that other frameworks of meaning, or value, or even consistent activity, can also provide structures and modes of inner organisation. It would be interesting to conduct studies on the effects of, say, the early learning of a musical instrument on ADHD, or of having an absorbing hobby, or simply of systematic learning or reading. But however concepts of meaning are conceived, the young self needs containing structures if it is not to fragment into atomised impulses, or mere receptors for stimulus and sensation. It needs to have ways of coping with time, and shaping it from within.

One telling and perfectly straightforward experiment, performed in Oregon by a psychologist named Michael Posner, showed that children with poor attention spans could be taught to concentrate by something like practice. In this study children were given video games which required alert concentration to be played well and which targeted specific aspects of attention. The games were repeated for five days, with children being supervised and encouraged to keep their attention focused. While video games may seem like a strange instrument for improving concentration, these were apparently designed just to this purpose. And surely it should come as no surprise that the children, even those with initially poor concentration and low levels of dopamine, began to show improved levels of attention and – yes – began to release more of the crucial neurotransmitter. [11]

Perhaps one thing that can be noted about this simple experiment is that the children participating in it were not only encouraged to focus their attention and extend it to longer durations; they were also *given* attention, of a focused and sustained kind. But this is exactly the kind of 'quality time' that many middle-class parents cannot afford to offer their progeny on a regular basis in their own over-busy, multitasking and segmented lives. Instead, the children's lives themselves are increasingly segmented as well, as they are ferried to their various improving activities or left to their own devices, in front of television or computer screens. In the meantime, those screens, and the technological surround within which children increasingly grow up, keep delivering aggressive and invasive versions of chopped-up time: moving images which can be made to flicker even faster by remote control, loud and shifting assortments of music, the super-fast and intrinsically violent kinetics of video games. This, too, must affect the very patterning of the young, not yet 'set' brain.

What is striking, of course, is how closely the symptoms of ADD mimic the patterns of contemporary, digital time. If 'the condition of immediacy' involves banishing the gap between demand and response, then children growing up under such a temporal regimen are unlikely ever to learn how to tolerate internal delay, or postponement of satisfaction. And it is hard not to notice that the list of symptoms attributed to this ailment closely corresponds to the routine symptoms of contemporary life: having one's attention scattered among countless stimuli, performing several tasks at once and thus not being able to give sustained attention to any one of them, demanding instant response from technological devices – the supposedly pathological manifestations of ADHD quite accurately describe our contemporary, middle-class modus vivendi in its everyday guise.

In that sense, the 'attention deficit' in ADD belongs not to the children but to the parents. In a book titled *The Tyranny of the Moment*, Thomas Hylland Eriksen, a Norwegian social anthropologist preoccupied with problems of time, notes, 'In the information society, the scarcest resource for people on the supply side of the economy is neither iron ore nor sacks of grain, but *the attention of others*.' [12] By 'people on the supply side of the economy' Eriksen means the various media which clamour for our attention, and among whose offerings we so insouciantly surf.

But the attention of others is becoming a scarce good in personal relationships as well, and the subjects who are most likely to suffer from its deprivation are the young. After all, the main framework of meaning, and source of emotional containment for children, is provided not by explicit values or ideas but by relationships with parents and intimates. As the adults in our cultures experience increasing amounts of stress and time anxiety, the children will undoubtedly pick this up and reflect it, not only on the behavioural but on the emotional and even physiological level.

Moreover, what the children are likely to pick up on are not only the daily schedules, but that inner organisation of time which evidences itself in the pace of movement, in the textures of nervousness or relaxation, in the quality of not only cognitive but emotional response. If the parents themselves flick from channel to channel, or from e-mail to computer games, the children will undoubtedly take on board not only similar habits but the chopped-up rhythms of such activity. If the habitual signals emitted by the adults around them are of speed and stress, the children will incorporate these less visible conditions into their own still-receptive selves.

Habitually anxious parents are likely to transmit anxiety to their children. If 'mirror neurons' are effective anywhere, it is surely in the transference of personal patterns to children, and such patterns extend to the shaping of both overt and internal time. The overt temporality of adults may well become the child's internal time; the unconscious disturbances of inner time in adults may manifest themselves in their offspring's overt symptoms.

In all these ways, the ADD generations are the legitimate and representative children of the digital epoch. This is increasingly acknowledged by various commentators, such as Richard Restak, a neuropsychiatrist who, in his book *The New Brain*, calls attention deficit 'the paradigmatic disorder of our times'. Indeed, so paradigmatic that, as Restak notes, many experts in the field are beginning to refer to ADD not as a disturbance, but as a cognitive style, or a 'distinctive type of brain organisation' adapted to the temporal conditions of our world.[13]

Perhaps so. But the segmentation and acceleration of time exacts its costs, above all, in children. Childhood is traditionally the stage of slow time – of exploratory play and errant curiosity, of sensuous wonder and even pleasant boredom. It is during the time in which nothing else happens that children can begin to take in the world around them, to imagine something which is not in front of them, to pick up an object in order to see how it is made, or to put something together in a new way. And it is in such time that children invent games to play with each other, thus exploring – without targets or schedules – modes of relationships and ways of being with others. Some of the most evocative descriptions of childhood time are to be found in Nabokov's *Speak, Memory* – both in the recollections of his own enchanted childhood and in the tender observations of his small son, over whom he watches as the boy stares endlessly at moving water, or performs repetitive manoeuvres with a toy car.[14] Adults could do worse in relation to children than heed Nabokov's injunction and warning: 'Never say hurry to a child.'

~

In response to fast time and fast technologies, we are developing 'fast' pathologies. The particular restlessness produced by the digitised version and vision of time – a vision of speed and stimulation for their own sake – manifests itself most clearly, or most neurotically, in syndromes like BPD or ADD. But in a less exacerbated, or perhaps less manifest form, the mental states and disturbances diagnosed in such disorders are experienced by many. Restlessness, running from one task to another or trying to do them all at once, running ahead of oneself and never catching up, this, too, is a common story. In turn, the discomforts caused by such endemic conditions breed their own reactions – and critiques.

In recent years there has been a growing literature on the contemporary problem of time: the *Faster* books by James Gleick; *In Praise of Slow* by Carl Honoré; *Time Wars* by Jeremy Rifkin; *Time: A User's Guide*, by Stefan Klein. Such books acknowledge the high level of time anxiety from which we are collectively suffering, and sometimes offer solutions, or at least alternatives. On one level there are the sociological analyses, diagnosing the time problem in measurable terms and addressing issues of what might be called temporal politics and policies. Concepts of 'hurry sickness' and 'time poverty' are becoming part of common sociological usage. And in at least one novel, *The Diagnosis*, by Alan Lightman, the protagonist goes mad from trying to squeeze his life into strictly scheduled modules of digital time.[15]

Such critiques themselves have a distinguished history. Here is Bertrand Russell, writing in the 1930s, in an essay called 'In Praise of Idleness': 'I want to say, in all seriousness,' the philosopher averred, 'that a great deal of harm is being done in the modern world by belief in the virtuousness of work, and that the road to happiness and prosperity lies in an organised diminution of work.'[16]

Russell insisted that he was not speaking from the position of an aristocratic dandy, and that he wanted his philosophy to apply equally to everyone. A four-hour working day, he thought, would be perfectly tenable for society as a whole, and the leisure hours gained thereby would allow everyone time for reading and thinking, for sport and self-cultivation, and for that idle speculation from which, he was convinced, most intellectual and creative discoveries emerge.

This was not, of course, to be, and perhaps Russell's hopes that everyone, in the right conditions, would use extensive leisure to cultivate temperate and philosophical pursuits were somewhat exaggerated. But the cause of idleness was taken up on a much larger scale in the 1960s 'counterculture' – that loose social movement which coincided with the coming of age of the baby-boom generation, and with its potential entry into the workforce in a period of unprecedented economic expansion. The counterculture was a multiheaded creature, but many of its manifestations can be seen as attempts to break away from the patterns of regulated time prevalent in the mainstream culture. The very concept of 'dropping out', which so centrally defined the 1960s, was a proclamation of protest against the imperatives of regular work, competitiveness and success – and, undoubtedly, against the anxieties these stimulated.

'Turn on, tune in, drop out.' The motto, coined by

Timothy Leary, one of the counterculture's prophets and gurus, for a while attracted quite a few acolytes, but the 'drop out' part of it turned out to be the least sustainable. For one thing, the very possibility of wholesale rebellion against the constraints of mainstream culture was underwritten by that culture's prosperity; for another, opting out of it involved giving up too many of its advantages. 'Dropping out' of a fast culture is different from living slowly in a slow culture. One cannot fully choose one's time schemes in isolation, any more than one can construct an entirely private language. Our constructions of time *are* to a large extent collective, and in order to participate in a culture, we have to accept some of its rules and patterns.

Moreover, Leary's siren call entailed an almost comical degree of temporal omnipotence. The guru himself urged his starry-eyed audiences not to trust anyone over thirty, rather ignoring the inevitable onset of that great age, even for his faithful followers. But the youth culture altogether was to a large extent based on a cult of its own youthfulness, and the illusion that it was possible to stay young for ever. 'Taaiiime is on my side – yes it is,' the Rolling Stones famously intoned; but that, of course, is never true for long.

The ethos of dropping out could not survive the sixties generation's entry, however delayed, into full adulthood. The baby boomers soon enough turned into a new bourgeoisie – working as hard as their parents, or in some cases harder. But the urge to escape the tyranny of 'mainstream' time has continued to be expressed in other, less wholesale ways. The vast drug subculture which arose during that period can be seen as one such expression. People take drugs for various reasons, of course, and to stimulate a variety of effects, but one of the chief attractions of chemical substances for their counterculture aficionados was their ability to alter the perceptions and sensations of time. The characteristic (or at least the most experimentally adventurous) drugs of the period were mescaline and LSD, and the characteristic effect produced by them was an extreme slowing down of inner time. This has a physiological and chemical basis; but for the sixties practitioners, the temporary abolishment of time was an important aspect of the hallucinogenic experience, and a crucial sign that 'the doors of perception' were being opened. Every formal and informal description of the 'high', or the 'trip' from that period contained paeans to the power of hallucinogens to distend the moment to vast proportions, to slow down the movement of time, or to make it stop and disappear altogether.

The drug subculture persists at the cultural margins into our own period, although these days the young turn to Ecstasy and other pill cocktails, while cocaine is the substance of choice for the upwardly mobile, and is as likely to be taken to boost performance in the hyper-speeded sectors of the financial professions as for the mystical experience of timelessness. There is also crystal meth, an easily available drug favoured by the fast-living and the hedonistic, which, in the words of one description, delivers a fast, 'cheap' sensation of pleasure – a 'rush' – by releasing dopamine and other hedonic substances into the brain. The immediate effects of crystal meth include increased alertness, excitement and rapid speech; the long-term effects can be devastating.

While drugs may these days mimic the symptoms of the culture rather than counteracting them, other forms of reaction against the pressures of 'mainstream time' have taken deeper root. The popularity of various spiritual and meditative movements, the proliferation of gurus, best-selling guides to spiritual health and weekend retreats – all of which have established themselves as such a central part of the cultural repertory in advanced countries – bespeak the need to find relief from digital time, to retreat to spaces which are not overcrowded with stimuli or over-packed with multiple tasks. Centres for meditation are now attended not only by the hip and the alternative, but by the harried and the professional looking for ways to calm down and restore scattered energies.

Whichever tradition they derive from, meditational practices address the problem of internal time through systematic techniques for regularising breathing, and through deliberate acts of turning attention inwards; that is, through both mind and body. The efficacy of meditation in slowing down internal pace is confirmed by recent studies such as that by Glicksohn noted in Chapter 2, which suggest that if we turn our minds away from the clock, or the urgent tasks awaiting us, and turn them inward instead, the neurological or subjective 'frames' within which time pulses become wider.[17] The pulse of time slows down. Cognitively oriented neuroscientists speculate that, at certain intense stages of meditation, the subject's attention is so fully turned towards internal states that the awareness of time disappears almost entirely. This may be scientific confirmation of the possibility of achieving cognitive timelessness (although it is not clear for how long this can be – or should be – sustained). But it should also be noted that such deployment of attention is the exact antithesis of the way we live now. Meditation in Hinduism has been described as 'a quiet, powerfully concentrated state wherein new knowledge and insights are awakened from within as awareness focuses one-pointedly on an object or specific line of thought'.[18] Aside from whatever seductions exoticism, esoteric religion or mantra yoga may offer, one can see the appeal of such concentration to the denizens of digital time.

Indeed, meditation is the perfect form of spirituality for our secular age – a practice which does not require specific religious belief but which draws on old disciplines for creating pockets of 'countertime'. The problem of time is historically central to all religions; indeed, it can be said that one of the main purposes of theological systems has been to provide answers to the menace and mystery of time. Broadly speaking, most traditional religions do this by dividing time into two separate domains: human and divine. Human time is the realm of change, decay and mortality; divine time is unmoving, changeless and eternal. The promise, and the consolation, for the believer in Judaism, or Christianity, or Islam, for example, is that after finishing our earthly span of existence we will be liberated into the domain of immutability, and of immortality; or at least, that such a domain exists, guaranteeing a larger, redemptive meaning for our brief span.

The premise, and promise, of meditative practices is different: what they offer are not visions of the afterlife, or of external eternity, but the possibility of achieving a sense of timelessness in the subjectivity, and in the here and now. It is this that the pressured professional craves, rather than visions of eternal salvation. And it is tempting to speculate that the enormous growth of the therapy culture in the 1960s, and its steady cultural presence since then, stems in part from similar needs. Therapy can sometimes address the malaises of modern temporality, but it also provides a temporary respite from its demands. Whatever else happens within the therapeutic hour, the hour is there, set aside for focused reflection, and not to be hurried or interrupted. The hour comes round (in principle) with unfailing regularity, and it guarantees (in principle) the regular and sustained attention of another person – that form of relational exchange whose deficit is so widely felt in contemporary lives. It is surely possible (and certainly, many psychotherapists would say so) that such temporal factors are responsible for a large part of therapy's therapeutic effect.

In addition to such broadly post-1960s phenomena, more recently western Europe has seen the emergence of modest but perhaps symptomatic groupings which address the issue of fast time not so much by creating alternative cultural spaces, as by slowing down the pace of ordinary activities. The slow food movement is the best known of these informal associations, and it has acquired something of a cultural status, with international adherents, institutional bases and conferences. But there are other manifestations of the slow-is-beautiful philosophy, with perhaps enough momentum to gather into not so much a counter- as an intercultural trend. In his book *In Praise of Slow*, Carl Honoré, a French journalist who travelled around Europe to collect his evidence, mentions a number of examples: slow cities, slow sex, slow childhood, and such clearly fey, if amusing, epiphenomena as a Bavarian-based Society for the Deceleration of Time.

Honoré discerns in such phenomena the beginnings of a 'worldwide movement' which is 'challenging the cult of speed'. But if there is a global groundswell of this rather gentle cultural current, it has not announced itself in evident ways. The 'slow' movements, with their temperate and rather sophisticated negotiations of fast and slow temporalities, have emerged mostly from advanced societies – that is, societies in which the acceleration of time has happened from within, and over the course of several decades. In other parts of the globe, the much more abrupt and sometimes externally imposed transitions to hypermodernity can lead to more radical and more reactionary (in the true sense of the word) responses. In the grievances of various orthodoxies against modernity, one can sense a revolt against the relentless speeding up of life, and the thoroughly utilitarian deployment of time, in which lived temporality becomes reduced to flat, segmented non-significance. The reversion to religious fundamentalism, so evident in various parts of the world, can be seen (on a very fundamental level) as an attempt to restore to time its dimensions and significance. The imperative to pray five times a day which obtains in Islamic religions, the injunction to keep the Sabbath in Judaism, the designation of certain days or periods as sacred – all these are ways of ordering time in the face of its hectic disarray. It is easy to understand the function and the appeal of such practices; it is possible to imagine the factory-hired Algerian peasants whom Bourdieu described – or their descendants – turning to fundamentalist Islam in defence against the extreme pressurisation, and the banality, of fast time. After all, the ritualisation of time was always a crucial aspect of religion, endowing sheer flux with symbolic markers and meanings.

But the import of religious temporalities is radically changed by their contemporary contexts. Post-modern religious orthodoxies create not only pockets of counter-time, but entire alternative temporalities – theologically sanctioned 'psychic retreats' within which the external time of the contemporary world can be ignored, or even denied. The attraction of such strategies is clear, but so are the dangers of psychic escapism, and its political twin, extremism. When the force of modern temporalities is evaded or disavowed, eventually the only way to meet its challenge and threat is to try to eliminate it. 'Today, ironically, the most virulent attempts to slow things down now take the form of national and religious fundamentalisms that deploy media sound bites and military campaigns of ethnic cleansing to return to a slow, centered world,' writes William E. Connolly in *Neuropolitics: Thinking, Culture, Speed*. 'The drives to pluralize and to fundamentalize culture form, therefore, two contending responses to late-modern acceleration.'[19]

It is possible that there are cultures in which there is no such thing as 'too slow'. Possibly the aboriginal people of Australia, following their songlines written in the earth and the movements of sun and shadow, never need hurry or press forward at anything but their natural pace. But in the contexts of modernity, and of societies which depend on more active constructions of time, slowness too easily becomes marginality, idleness comes close to uselessness. One need only think of the perpetually reclining hero of *Oblomov*, that classic novel of fortune-funded indolence by Ivan Goncharov, or the luxuriously unoccupied aristocrats of Chekhov's *Uncle Vanya* and the sense of aimlessness which accompanied their condition. The Russian concept of the 'superfluous man' came from this upscale form of unemployment. For bourgeois women, for whom it would have been shameful to work, there were long days spent in conspicuous inactivity. Stifled heroines such as Flaubert's Emma Bovary are, on one level, representations of enforced idleness and the sheer boredom it brings – with all the restlessness or recklessness or penchant for melodrama that can follow from this seemingly undramatic state.

In the context of contemporary acceleration, projects

aiming for a return to earlier forms of temporality become even more untenable, calling for a sort of collective equivalent of psychic regression, or at least of living in the past. Moreover, the costs of reactive slowness in productivity and economic competitiveness are too high. As Connolly notes, 'In today's world it is less that the large consume the small, and more that fast process overwhelms slow activity.'[20] In a world riven by 'asymmetries of pace', fast time trumps slow time every time.

~

Slow may not be always beautiful, and fast has its seductions and its pleasures. In collective terms, speed is competitive. On the individual level it bespeaks energy, vitality, alertness of body and of mind. We admire the streamlined movement of the forward-stretched runner, the quickness of reflex and response, the invisible agility of mental operations.

Fast technologies have taken over the world partly because they are attractive, and few of us would willingly give up the benefits they bring: simplification of many activities, variety of stimulus, global communicative reach. Speed sometimes equals spontaneity, and the instantaneity of digital operations may release certain energies or styles of thought which have their own appealing and interesting qualities. E-mail may encourage grammatical laxness or risky impulsiveness, but it can also make for a liveliness of expression, a certain attractive freshness, telegraphic condensation and communicative immediacy which are in some ways reminiscent of those rapid and frequent notes and *billets doux* which travelled between enemies and lovers when couriers in large cities delivered mail several times a day. And while texting often seems not so much illiterate as sub-linguistic, it may be creating elements of a new vernacular, with its own possibilities for inventiveness or even expressiveness.

It is also possible that instant access to diverse sources of information encourages a multivalent mode of perception which corresponds more closely to our interconnected world, with its intermingled and multiple perspectives. When we receive news from all parts of the globe simultaneously, when we conduct business transactions or personal communications across cultures and time zones, it becomes more difficult to conceive of ourselves as the only locus of importance, or even a privileged centre of vision. We know, palpably and concretely, that there are numerous loci of activity and points of view to take into account.

And yet, both in the social realm and on the individual level, fast time has many unintended and much less beneficial consequences. Indeed, the multiplication of unintended consequences is one predictable consequence of hyperspeeds operating at a massive scale. In the nanosecond temporality of our technological functions we have far surpassed the velocities of industrial time, and the multidirectional, simultaneous flows of data with multidimensional and unprocessed significations have the potential to produce unforeseen and uncontrollable effects: collisions, communicative gridlock, or the collective inability to analyse the information being produced and delivered in such incalculable quantities. Market traders standing in front of bourse screens, their hands clapped to their heads in dismay as stock values leap or plunge seemingly without human intervention, are one iconic image of our time. The global market crisis unfolding at the time of this essay's writing can be seen at least in part as a consequence of cyberspace speeds: economic chaos produced by criss-crossing interactions of virtual data so incessant, instantaneous and self-multiplying as to defy control or proper analysis. The *New York Times* talks of 'a web of risk,' and 'virally connected' economic transactions which slipped out of anyone's control. Other commentators warn of longer-range risks of digital time. The French cultural theorist Paul Virilio has written about the dangers inherent in the escalating speeds of military technology, with the resulting potential for impulsive, remote-control actions (just push that button as you see some enemy planes coming in the wrong direction) which are abstract in themselves but can have deadly repercussions – the Dr Strangelove scenario establishing itself as a routine part of war. [21] Sociologists and political analysts worry that the pace of our cultural and technological developments is incompatible with what the noted American political scientist Sheldon Wolin calls 'deliberative democracy' – a form of governance which requires its own, much slower pace of negotiation, public discussion and decision making.[22]

Another French theorist, Jean Baudrillard, takes this

idea further, suggesting that in hurtling along on the currents of sound bites, buzz and instantly forgotten information, we are losing the very notions of historical time, or 'event'.[23] In order for disparate occurrences in the political space to coagulate into an 'event', we need to make connections between them, to reflect on their meanings, and to see their shape. Instead, as indiscriminate information keeps bombarding us from all sides, the happenings in the public realm remain just that: something that happens, and then disappears. Journalists and bloggers produce information in real time; we receive it in a constant succession of disconnected data. And, in this perpetually created and perpetually vanishing present, there is no time to take stock, to consider relationships between occurrences, or link into the longer and deeper time of history.

The pessimism of such sociocultural arguments can be, and has been, contested. The counterarguments point out that the web offers new possibilities for political participation and expression to people who would not otherwise have access to public fora. Small and marginal groups can communicate with like-minded others who are concerned with similar issues; people under oppressive regimes can receive information from other parts of the world which is censored in their own countries. That, after all, is why governments in various authoritarian states try to prohibit or limit access to the internet.

The sociopolitical impacts of mass-scale speed are complex. But it is on the individual level that the repercussions of fast time are most immediately and most deeply felt – and it is in our ordinary lives that, in a sense, they most matter. What happens to our minds and psyches when we habitually operate in a cultural/technological environment in which time has been abstractly miniaturised to the nanosecond, and experientially reduced to the instant and the condition of immediacy? What are the consequences of our contemporary constructions of time for our lived experience and our inner lives?

To try to grasp paradigmatic change as it happens is perhaps to wander where angels fear to tread. But, to hazard such a step, it seems to me that what may be at stake is the very dimension of subjectivity, or inwardness – a topological space so seemingly intrinsic to the self as to be inseparable from the idea of the human; but whose textured dimensionality may be jeopardised by digital time in various ways. On one level we are relegating more and more of our mental operations to various technologies, with digital devices increasingly acting as prostheses for our faculties. We entrust our sense of spatial orientation to satellite navigation systems; we give mathematical calculations over to the appropriate gadgets. Indeed, the temptation is to let the computer do much of our thinking for us. We can cut and paste fragments from the internet and hope that the collage adds up to something coherent, or interestingly disjunctive. Certainly, we have less need to remember information ourselves when so much can be stored in the computer's memory. The feats of memory recorded in oral cultures, or performed by Soviet poets and writers under censorship, seem hardly credible within our zeitgeist. Nadezhda

Mandelstam memorised all of her husband's poetry because it was too hazardous to write it down. Solzhenitsyn committed to memory each page he wrote when he was imprisoned in the Gulag, and then destroyed the evidence. Such powers of retention are unimaginable to most of us and they may become even more so, as we transfer memory to the many storage places available to us – there to be filed away, for instant and effortless retrieval.

By transposing aspects of thought and memory to technology we are externalising our mental operations. But both the mind and the psyche require internality. In order to reflect on a problem or build an argument, we need to turn mental attention inwards, to mull over ideas and let the mind wander; to sift the important from the trivial; to follow thoughts in their course and consider the disjunctions and connections between them. If the occurrence of a single thought happens in measurable time, then the development of ideas, or the elaboration of a creative impulse, or self-examination, all happen in more extended durations, and need more persistent concentration. Acts of mind require us, at least occasionally, to be unplugged. ('Only disconnect' was a motto I was briefly tempted to adopt; but that, I realised, was counsel of despair.)

And if we want to make sense of our own affective experience, we also need to dip inwards and travel through our subjectivity; to 'work through' the content of our feelings and let them sift through strata of the self; to weave disparate episodes of our lives and interpret and reinterpret their meanings. Whatever the level of our cognitive memory, or capacity to retain data, the registration and encoding of long-term memory requires literal time; the conversion of short-term into long-term memory takes at least twenty-four hours. But the transmutation of discrete memories into personal remembrance, with all its reverberations and resonances, surely requires not only more but another *kind* of time. In order to make meaning of our memories (and the experiences from which they flow), we need to give them our affective attention, to filter them through the scrims of our own subjectivity and absorb them into the networks of memories and selves we already are. Freud spoke of 'the work of mourning', but there is also the work of memory and of affective reflection, and it requires us to descend, or peer into, our internal spaces and the elusive dimension of inner depth.

On hearing the bell which brings back the panorama of his life at the end of his great opus, Proust writes, 'for in order to get nearer to the sound of the bell and to hear it better it was into my own depths that I had to re-descend. And this could only be because its peal had always been there, inside me, and not this sound only but also, between that distant moment and the present one, unrolled in all its vast length, the whole of that past which I was not aware that I carried within me.'[24]

The processes of reflection, and of subjectivity, take time – that peculiar, non-mathematical temporality which moves at its own errant pace and with its own meandering rhythms. Inwardness cannot be submitted to the calculations of efficiency or to the reign of immediacy. Experience does not happen in instants. Neither do relationships of any but the most instrumental kind.

The arena of relationships – forms of friendship and of intimacy, modes of meeting others and of knowing them – is among the areas of life being most saliently changed by contemporary technologies. Among the generations who have grown up with the computer, social relationships are conducted largely via the internet, and their numerical possibilities at least are increased geometrically by the multiplicative powers of that instrument. The bulletins posted on Facebook can be sent with frequency and facility to hundreds of people simultaneously. Undoubtedly, the ease and reach of such modes of communication can yield interesting and manifold opportunities and encounters. But one wonders what, by the emphasis on multiplicity and simultaneity in this domain as well as in others, is lost in those imponderables of depth and intensity. One recent convert to Facebook (quoted in the *New York Times*) avers that, by sending frequent bulletins to many others as she moves through her days – 'I am stuck in a traffic jam', 'I have a headache coming on', 'I just bought a great new dress', – she feels not only that she knows more people, but that she is better known. But this raises the question of what kind of knowing that is. The impersonal, indeed virtual, gaze of scattered interlocutors hypothetically noting fragmented bits of one's activities is surely not the same as the embodied attention of another person, taking in what you are saying and following the thread of your story. It is not the same as the gaze which in some way witnesses your life.

The accumulation of knowledge on the internet involves what is sometimes called 'stacking' – placing different elements of information next to each other or on top of each other without modifying any of them through the juxtaposition. But intersubjectivity, like subjectivity, requires precisely a deepening and development of knowledge rather than its mere accumulation; it requires the gradual growth of understanding, the sedimentation of new perceptions and feelings in the mind and the body until they become part of the fabric, or the material, of the self. The domain of relationships, above all, is one in which quantity does not equal quality, and speed is no guarantee of satisfaction. Coming to know another person calls for a certain affective energy and sustained attention; for the willingness to travel into the inwardness of another person, to probe behind appearances, to let empathy follow its own unpredictable temporal pathways. Intimacy can rarely fit into sound-bite intervals, and it rarely happens on schedule.

The cost of conducting relationships via communication technologies is that they subsume all human exchanges to the modular, segmented temporality of their own operations. At another scale of experience, our habitual and sometimes relentless geographical mobility fragments the *longue durée* of lived time, and with it the textures of personal relations and other attachments.

Excessive stability has its problems. The over-regulated bourgeois households familiar from Mann's

Buddenbrooks or, for that matter, Sartre's memoir *The Words*, with their unvarying routines and slow-ticking grandfather clocks, were, for some, stifling in their boredom. Sartre's breakout into devil-may-care freedom and the existentialist construction of time – each act authentically chosen, each act new – was propelled in part by a rebellion against his grandfather's stolid stability (based on an unshakeable belief in steady progress) and equally stolid self.

But it is one comfort (as well as a possible discomfort) of rooted lives, spent in a village, say, or a small locality, that the trajectory of life's stages unfolds with a certain continuous, or at least visible, logic; one can witness the lives of others from the beginning to the end and the development of one's own history can be similarly witnessed and followed.

Rootedness, however, is becoming the exceptional rather than the usual circumstance. Our normative conditions comprise global mobility and the routine nomadism of travel. The phenomena of exile, emigration and cross-border movement have become a central feature of our period. And, of course, movement in itself, like speed, can be energising. Mobility is a form of freedom; it is born of options and can offer second chances, new beginnings, possibilities for self-reinvention. And yet there are costs, known to great numbers of immigrants, refugees and professional nomads circling our globe: fees to be paid in the breakage of cultural time, loss of internal continuity, the disruption of one's personal world. Milan Kundera's *The Unbearable Lightness of Being* has become the urtext (and phrase) of this condition. Sabina, its 'light' protagonist, chooses exile over native land, movement over fixity, detachment over sentiment. But with it she chooses a freedom which eventually becomes an unbearable burden, and a poignant loneliness in which nothing and no one in her surroundings really matters.

We can shuttle among parts of the globe, shedding problems, and sometimes identities, as we go along. In narrative terms, our lives are becoming more picaresque than nineteenth-century realist. The new self, no less than the new brain, is likely to be characterised by lightness rather than gravitas, irresponsibility rather than weighty conscience, by mercurial changeability rather than the solidities of 'character'. Sociologists and cultural commentators have been tracking such developments for a while. Robert J. Lifton's 'Protean self', Zygmunt Bauman's 'liquid modernity', or the concept of 'modernity lite' are all attempts to capture the lighter and more fluid forms of selfhood created by contemporary cultural conditions and new kinds of movement.[25] Motion creates time, and forms of movement construct lived temporality. Habitual mobility liberates us from the responsibilities of linear time, and the weighty patterns of cause and consequence. But it also deprives us of certain intensities of attachment. It means we can choose new sites of life at will, but also that we know them less well, and are less fully known.

~

There may be little to be nostalgic for in the Puritan ethic, or the heavy monotonies of bourgeois lives we are emerging from. But monotony is hardly our problem today. With such wide availability of instant information and entertainment, we need never be bored or have time weigh heavily on our hands. Our cultural ego-ideals include multiplicity, stimulation, instant gratification. We believe in the glamour of speed, in the sexiness of multitasking, in the survival of the fastest.

On another level, our cultural attitudes towards time are characterised by what might be called temporal omnipotence. At one end of the spectrum, we believe we can extend time indefinitely by living longer, staying younger and – who knows – perhaps fending off mortality altogether. At the micro scale, we try to compress each moment by loading it with maximum stimulus and activity. We want more: more lifespan, more experience packed into every moment, more speed per unit of experience, more places compressed into the time we have, more time zones compressed into the locality in which we are physically placed. We believe we can manipulate time at will.

In a sense, we still hew to an instrumental, late-capitalist version of Chronos: we try to exploit every moment for all it is worth, and to squeeze out of it maximum value. But this is no longer wedded to a teleological vision of time, or the ideology of progress, or even the aspiration to cumulative achievement. There are good reasons to suppose that the anxieties of post-modernity – the states most clearly typified in disorders such as BPD or ADD

– are different in kind and texture from the sufferings brought on by the repressions of the Protestant ethic. The peculiar disquietude of our time is not fuelled by the conviction that it is sinful to waste time, or the guilt for giving in to a transgressive impulse, in lieu of postponing pleasure; rather, it is propelled by the lack of an overarching temporal structure on the one hand and the hedonism of the present on the other.

It is as if we have substituted speed for significance. If we move fast enough, we can fill up time without reflecting on our intentions or purposes. Speed becomes its own self-justifying value. But the decoupling of activity from larger frameworks of meaning, or a sustaining temporal vision, carries its own anxieties. It fragments time internally, as well as externally; for without some sense that disparate occurrences are linked through coherent meanings or aims, there is no way to connect the moments in which they occur. In such a construction of time, the pressure on each moment increases enormously, creating a heightened state of internal pressure; the hedonism of immediacy leads to its own peculiar discontent.

This is the basic paradox: if we try to pack too many experiences into our time, then we lose the ability to experience – to process occurrences through the filter of our selves until they become ours, and part of us. The human organism functions at different temporal scales: the micro units of cognition, the daily cycles of the body, the rhythms of life-stages and the variable durations of the psyche. At each level we shape time, and at each level human temporality has its parameters and limits which cannot be violated without some hazard. Is it possible that on the most important level of lived, whole experience there are also unwritten temporal laws and norms?

Excessive deceleration is clearly not the answer to excessive acceleration. Slow may be desirable sometimes, and idleness next to godliness in the right measure. But immoderate languor can also lead to a disorganisation of time and loss of temporal form. On the physiological level the retardation of metabolic processes beyond a certain point results in unconsciousness or death. And in the psyche as well, immobility is a sign, and sometimes the cause, of depression or hopelessness or the flight from vitality and change. Energies which are unused become sour and turn in destructively on the self. Time which is insufficiently filled with activity dissolves into disorienting amorphousness.

In *Flow: The Psychology of Happiness*, Mihaly Csikszentmihalyi, an American social psychologist, takes a different tack on the question of proper experiential pace by examining the kinds of activities – and deployments of time – which he and his many interviewees consider most conducive to pleasure and enjoyment.[26] Such activities are found in many areas of life: games, sports, dancing, creative endeavour, or even perfectly ordinary jobs – but they all must meet several conditions. They have to be demanding, but not impossible to accomplish; they have to have well-defined goals and be, to the person who performs them, purposeful and self-justifying. Such frameworks of activity produce what Csikszentmihalyi calls 'optimal experience': states of focus, concentration, total engagement in activity and, above all, flow: that is, a sense that time is moving at the right, unforced pace.

It is interesting that most up-to-date observers deploy the most ancient metaphor – flow – to describe the nature of time when it feels most natural. Time flows. And it is when we can plunge into its flow and move through it without excessive resistance or excessive strain, that we can attain a sense of enjoyment and gratification. Apparently, satisfaction also happens in its own *tempo giusto*; neither too fast, nor too slow.

~

We are, in astonishing ways, temporal creatures. Our bodies are their own timers; we take in time through our minds and senses, and we create time shapes within us. But beyond that, the problem of time is inseparable from that of meaning. Time *is* the fundamental medium and condition of human meanings. It is the finitude of that element which is the ground of all existential quandaries. We think against the horizon of mortality, and contemplate first questions against the knowledge of our own ending. What are the uses of a finite life, and what uses do we want to make of it?

Some of the richest meditations on such questions have always been found in art – especially the time-extended media of literature, film and music. It can be said that time is the metaphysical ether of art, giving it substance and breath. The passage of time, with all its poignancies, is the express theme of many great works; but even when it is not, temporality is often the subterranean subject, revealed through structure and form. This is especially true of music, which, of all the arts, works most directly with time, both as its subject and material. In a sense, music is nothing but the shaping of time, and the constructions of time it is capable of evoking are infinitely varied. Musical motion can combine simultaneity with extension, extreme speed with extreme slowness, motoric regularity with the graceful swoop of feeling, the pointillism of separate moments (and notes) with the movement (in melody and the longer arc of form) from the past to the present; music can express the agitated conflicts of intense emotions, and the calmer gestures of acceptance and reconciliation. Complex compositions can map, and represent, more complex topologies as well, including a self-reflective interpretation of time, as motifs are repeated in different contexts and their implications, or valences, are modified in the reiteration.

Moreover, the structures of music can invoke what used to be called the zeitgeist, or the spirit of the times, and it is interesting that, after a period of modernist fragmentation, contemporary composers such as Steve Reich and Philip Glass are experimenting with forms which seem to combine the regularities of Bach with some very post-modern intuitions and patterns. These composers' techniques of reiterating, with minute variations, short musical units, can be seen as both a way of slowing down time and of registering certain processes which may take place both in computers and in our bodies – the repetition, replication and recombination of microscopic, modular units of matter and motion.

But it is the power of music, of all periods, that it has literally therapeutic properties. Music can help alleviate symptoms of physical illness (for example, in Parkinsons patients), and is used to diminish the pain of psychic disorders such as autism or, indeed, ADD. Patients with severe memory loss often remember music after they have forgotten everything else. Possibly, this is so because the constructions of time in music correspond to the temporal structures of the body and psyche.

Music is the medium most capable of expressing all aspects and dimensions of both measurable external time and subjective lived temporality. And it is the profound effect music has on us that most clearly demonstrates our deep need for meaningfully ordered time. But perhaps also the musical moulding of time can point the way towards beneficial transactions with that element, as we enter the rapidly approaching future. We need to structure time and fill it with meaning. But in our complex, demanding, multidimensional lives we may also need to acknowledge that heteronomy of time – the fluctuating variety of rhythms, motions and emotions – which is found in music, but also in subjectivity.

Fast time has its allure, and its liberating possibilities. The Greeks acknowledged a sort of ontological heterechrony by naming two kinds and concepts of time: Chronos, the time of continuity and mutability, and Kairos, the temporality of the auspicious moment, of opportunity or crisis – the kind of heightened and irretrievable instant that we need to grab by the horns, or the head. In a less philosophical vein, there are some decisions we can make in the blink of an eye, and some tasks we want to accomplish as efficiently as possible. Retrieving money from an ATM beats standing in a bank queue any day. Technology can give us unprecedented flexibility and freedom. We can perform routine or tedious tasks anywhere, and often at times of our own choosing.

But if we subsume all our activities to the categories of instrumental time, we run the risk of losing dimensions of experience which are among the richest parts of the human repertory. In *Slowness*, Kundera evokes the sensuality of actual seduction, as it unfolds with the languorous and stately pace of a minuet. But in order to feel the enchantment of the world altogether, we need sometimes to give ourselves to its seductions with a kind of receptivity, rather than headlong forward determination. We need intervals in which to savour and relish experience rather than rushing through it, or hitting the keypad. Nabokov (my fellow traveller throughout this book) felt his first poem arise in his mind as he watched a water droplet glide down a leaf after a heavy shower; a pause in which time stopped and expanded, and then regained, in the poem, its flow.

We do not all have to be poets, but if we do not want to live meaninglessly, then we need to give ourselves over sometimes to the time of inwardness and contemplation, to empathy and aesthetic wonder. We need to mull and muse, to reflect on our experience and interpret it, to perform on the level of our life narratives those acts of autopoiesis which apparently happen outside our intention or ken in the brain's neurological pathways. We need occasionally to go with the flow.

Temporal omnipotence is the most omnipotent form of omnipotence, for it tries to defy the inexorable and the inevitable. Time places a limiting condition not only on our lifespan but on the ideology of will, and the illusion of total control. In a thoughtful book entitled *Timescapes of Modernity: The Environment and Invisible Hazards*, Barbara Adam argues that the depredations we have visited on the environment have followed from our denial of nature's temporality – the hidden, slow processes of growth and decay, the long-term effects of oil spills or deforestation.[27] We know the damage our exploitations of nature have wrought. But if we over-exploit our own, human temporality we risk similarly tearing and degrading the experiential fabric. Technology and new knowledge can bring us more health and longer lives, more options and more instant gratifications. But if we use ourselves instrumentally and do not cultivate the subterranean strata of the self, if we do not take care because we believe we will live for ever, and if we try to pack all moments with digital quanta, then we run the danger of laying waste, or killing the time that is given to us. In the Faustian bargains we try to strike with time, that would be the most ironic price of all. At least, from my admittedly anthropomorphic perspective, so it seems to me. The workings of time in ourselves are imperceptible, but we need to recognise and respect both its parameters and our own temporal limits if, in our rush to post-natural modes of existence, we do not want to lose a rich and deep measure of our humanity.

NOTES

Introduction

1. Vladimir Nabokov, *Speak, Memory: An Autobiography Revisited* (Penguin Modern Classics, 2000).
2. Carmen Firan, 'Inside and Beyond Words', *Aspasia*, vol. 2, no. 1 (March 2008), p. 197.
3. Ibid.
4. Jeremy Rifkin, *Time Wars: The Primary Conflict in Human History* (Henry Holt, 1987), p. 5.

1 Time and the Body

1. Stephen Hawking, *A Brief History of Time: From the Big Bang to Black Holes* (Bantam, 1995).
2. Jonathan D. H. Smith, 'Time in Biology and Physics', available at http://www.chronos.msu.ru/EREPORTS/smith_time.pdf, p. 6.
3. *Thomas Nagel, 'What Is it Like to Be a Bat?', Philosophical Review, vol.* 83 (1974), pp. 435–50.
4. Jon Mooallem, 'The Sleep-Industrial Complex', *New York Times*, 18 November 2007.
5. Russell G. Foster and Katharina Wulff, 'The Rhythms of Rest and Excess', *Nature Reviews Neuroscience*, vol. 6 (May 2005), pp. 407–14.

6. Jeanette Winterson, 'Disappearance I', in Winterson, *The World and Other Places* (Vintage, 2000).
7. Frank J. Tipler, *The Physics of Immortality* (Doubleday, 1994), p. iii.
8. André Klarsfeld and Frédéric Revah, *The Biology of Death: Origins of Mortality*, trans Lydia Brady (Cornell University Press, 2003).
9. Robert N. Butler, *The Longevity Revolution: The Benefits and Challenges of Living a Long Life* (PublicAffairs, 2008).
10. Simone de Beauvoir, *All Men are Mortal*, trans. Leonard M. Friedman (W. W. Norton, 1992), pp. 172–3, 339.
11. Julian Barnes, *Nothing to be Frightened Of* (Cape, 2008).
12. Peter W. Hochachka and Michael Guppy, *Metabolic Arrest and the Control of Biological Time* (Harvard University Press, 1987), p. 1.

2 *Time and the Mind*

1. Gerald Edelman, *Wider than the Sky: A Revolutionary View of Consciousness* (Penguin, 2005), pp. 102–3.
2. Ibid., p. 55.
3. Quoted in Israel Rosenfield and Edward Ziff , review of *The Physiology of Truth: Neuroscience and Human Knowledge* by Jean-Pierre Changeux, *New York Review of Books*, 26 June 2008.

4. Richard Restak, *The New Brain: How the Modern Age Is Rewiring Your Mind* (Rodale International, 2004), pp. 56–7.
5. Joseph Glicksohn, 'Temporal Cognition and the Phenomenology of Time: A Multiplicative Function for Apparent Duration', *Consciousness and Cognition*, vol. 10 (2001), pp. 1–25.
6. For example Bernard Pachoud, 'The Teleological Dimension of Perceptual and Motor Intentionality', in Jean Petitot, Francisco J. Varela, Bernard Pachoud and Jean-Michel Roy, eds, *Naturalizing Phenomenology: Issues in Contemporary Phenomenology and Cognitive Science* (Stanford University Press, 1999), ch. 6.
7. Oliver Sacks, 'The Lost Mariner', in Sacks, *The Man Who Mistook His Wife for a Hat: And Other Clinical Tales* (Touchstone, 1998), pp. 23–42.
8. Ibid., p. 30.
9. Walter Kirn, 'The Autumn of the Multitaskers', *Atlantic Monthly*, November 2007, p. 72.
10. Ibid., pp. 66–75.
11. Antonio Damasio, *The Feeling of What Happens: Body, Emotion and the Making of Consciousness* (Heinemann, 2000).
12. Franz Kafka, '*Reflections on Sin, Pain, Hope, and the True Way*', in *The Great Wall of China*, trans. Willa Muir and Edwin Muir (Schocken, 1946).
13. Jean Piaget, *The Child's Conception of Time* (Routledge, 2006).

14. Adrian Johnston, *Time Driven: Metapsychology and the Splitting of the Drive* (Northwestern University Press, 2005), p. 3.
15. Sigmund Freud, 'The Structure of the Unconscious', in Freud, *An Outline of Psychoanalysis*, trans. James Strachey (W. W. Norton, 1949).
16. David Bell, 'Existence in Time: Development or Catastrophe?', in Rosine Jozef Perelberg, ed., *Time and Memory: The Power of the Repetition Compulsion* (Karnac, 2007), pp. 65–84 at p. 65.
17. Paul Williams, 'Making Time, Killing Time', in Perelberg, ed., *Time and Memory*, pp. 47–64 at p. 52.
18. Matthew R. Broome, 'Suffering and Eternal Recurrence of the Same: The Neuroscience, Psychopathology, and Philosophy of Time', *Philosophy, Psychiatry, & Psychology*, vol. 12, no. 3 (2005), pp. 187–94.
19. Thomas Fuchs, 'Fragmented Selves: Temporality and Identity in Borderline Personality Disorder', *Psychopathology*, vol. 40, no. 6 (2007), pp. 379–87 at p. 379.
20. Ibid., p. 386.
21. Kim A. Dawson, 'Temporal Organization of the Brain: Neurocognitive Mechanisms and Clinical Implications', *Brain and Cognition*, vol. 54 (2004), pp. 75–94.
22. Jacques Lacan, *Ecrits: A Selection*, trans. Alan Sheridan (Routledge, 1990).
23. André Green, *Time in Psychoanalysis: Some Contradictory Aspects*, trans. Andrew Weller (Free Association Books, 2002).

24. Adam Phillips, *Darwin's Worms* (Faber and Faber, 1999), p. 26.
25. Williams, 'Making Time, Killing Time', p. 55.

3 Time and Culture

1. Ryszard Kapuscinski, *The Shadow of the Sun*, trans. Klara Glowczewska (Vintage, 2002), p. 17.
2. Robert V. Levine, *A Geography of Time: The Temporal Misadventures of a Social Psychologist, or How Every Culture Keeps Time Just a Little Bit Differently* (Basic Books, 1998), pp. 159–60.
3. Clifford Geertz, *The Interpretation of Cultures* (Basic Books, 1997), p. 389.
4. Ibid., p. 391.
5. Ibid., p. 393.
6. Jerome Bruner, 'Life as Narrative', *Social Research*, vol. 71, no. 3 (Fall 2004). See also *Acts of Meaning: Four Lectures on Mind and Culture* (Jerusalem-Harvard Lectures) (Harvard University Press, 1992).
7. Pierre Bourdieu, 'The Attitude of the Algerian Peasant Toward Time', in Julian Pitt-Rivers, ed., *Mediterranean Countrymen* (Mouton, 1963), pp. 52–77 at pp. 57–8.
8. Ibid., p. 71.

4 Time in Our Time

1. Arlie Russell Hochschild, *The Time Bind: When Work Becomes Home and Home Becomes Work* (Holt Paperbacks, 2001).
2. Ibid., p. 46.
3. Leon Kreitzman, *The 24 Hour Society* (Profile, 1999), p. 25.
4. Cited ibid., p. viii.
5. Robert E. Goodin, James Mahmud Rice, Antti Parpo and Lina Eriksson, *Discretionary Time: A New Measure of Freedom* (Cambridge University Press, 2008).
6. Wim Veen and Ben Vrakking, *Homo Zappiens: Growing Up in a Digital Age* (Continuum, 2006).
7. John Tomlinson, *The Culture of Speed: The Coming of Immediacy* (Sage, 2007), p. 91.
8. Larry R. Squire et al., eds, *Fundamental Neuroscience* (Elsevier Science, 2002).
9. See Eric Kandel, *Psychology, Psychoanalysis and the New Biology of Mind* (American Psychiatric Publishing, 2005).
10. For example Marvin Zuckerman, 'The Neurobiology of Impulsive Sensation Seeking: Genetics, Brain Physiology, Biochemistry and Neurology', in Con Stough, ed., *Neurobiology of Exceptionality* (Kluwer Academic/Plenum Publishers, 2005).
11. Michael Posner et al., 'Training, Maturation, and Genetic Influences on the Development of Execu-

tive Attention', *Proceedings of the National Academy of Sciences*, vol. 102, no. 41 (2005), pp. 14931–6.

12. Thomas Hylland Eriksen, *The Tyranny of the Moment: Fast and Slow Time in the Information Age* (Pluto, 2001), p. 21.
13. Richard Restak, *The New Brain: How the Modern Age Is Rewiring Your Mind* (Rodale International, 2004), pp. 47, 46.
14. Vladimir Nabokov, *Speak, Memory: An Autobiography Revisited* (Penguin Modern Classics, 2000).
15. James Gleick, *Faster: The Acceleration of Just About Everything* (Vintage, 2000) and *Faster: Our Race Against Time* (Abacus, 2005); Carl Honoré, *In Praise of Slow: How a Worldwide Movement Is Challenging the Cult of Speed* (Orion, 2005); Jeremy Rifkin, *Time Wars: The Primary Conflict in Human History* (Henry Holt, 1987); Stefan Klein, *Time: A User's Guide* (Penguin, 2008); Alan Lightman, *The Diagnosis: A Novel* (Vintage, 2002).
16. Bertrand Russell, *In Praise of Idleness: And Other Essays* (Routledge Classics, 2004), p. 3.
17. Joseph Glicksohn, 'Temporal Cognition and the Phenomenology of Time: A Multiplicative Function for Apparent Duration', *Consciousness and Cognition*, vol. 10 (2001), pp. 1–25.
18. As defined by the Himalayan Academy, a Hawaii-based Hindu community.
19. William E. Connolly, *Neuropolitics: Thinking, Culture, Speed* (University of Minnesota Press, 2002), p. 180.
20. Ibid., p. 143.

21. Paul Virilio, *The Information Bomb*, trans. Chris Turner (Verso, 2000).
22. See William E. Connolly, *Neuropolitics: Thinking, Culture, Speed* (University of Minnesota Press, 2002), pp. 35–6 and chapter 6: 'Democracy and Time', pp. 140–73.
23. Baudrillard, Jean, *The Illusion of the End*, trans. Chris Turner (Polity Press, 1994).
24. Marcel Proust, *Remembrance of Things Past*, trans. C. K. Scott Moncrieff and Terence Kilmartin (Vintage, 1982), vol. 3, p. 1105.
25. Robert Jay Lifton, *The Protean Self: Human Resilience in an Age of Fragmentation* (University of Chicago Press, 1993); Zygmunt Bauman, *Liquid Modernity* (Polity Press, 2000).
26. Mihaly Csikszentmihalyi, *Flow: The Psychology of Happiness* (Rider, 1992).
27. Barbara Adam, *Timescapes of Modernity: The Environment and Invisible Hazards* (Routledge, 1998).

BIBLIOGRAPHY

A volume on a subject as multidisciplinary as time is necessarily in part a work of synthesis, drawing on primary research in many fields. I gratefully acknowledge my debt to such exploratory work. What follows is a list of sources and background reading which – among others – have informed this book.

Adam, Barbara, *Timescapes of Modernity: The Environment and Invisible Hazards* (London and New York: Routledge, 1998)

Ashley, Jackie, 'Feeling Time-Short Is Not Just a Soft and Fluffy Issue', *Guardian*, 18 June 2007, p. 33

Barnes, Julian, *Nothing to be Frightened Of* (London: Jonathan Cape, 2008)

Baudrillard, Jean, *The Illusion of the End*, trans. Chris Turner (Cambridge: Polity Press, 1994)

Bauman, Zygmunt, *Liquid Modernity* (Cambridge: Polity Press, 2000)

Beauvoir, Simone de, *All Men Are Mortal*, trans. Leonard M. Friedman (New York and London: W. W. Norton, 1992)

Behnke, John A., *The Biology of Aging* (New York: Kluwer Academic/Plenum Publishers, 1978)

Bell, David, 'Existence in Time: Development or Catastrophe?', in Rosine Jozef Perelberg, ed., *Time and Memory: The Power of the Repetition Compulsion* (London: Karnac, 2007), pp. 65–84

Ben-Shahar, Tal, 'Cheer Up. Here's How . . .' *Guardian*, 29 December 2007, p. 36

Bergson, Henri, *Time and Free Will* (London: Elibron, 2005)

Bourdieu, Pierre, 'The Attitude of the Algerian Peasant Toward Time', in Julian Pitt-Rivers, ed., *Mediterranean Countrymen* (Paris: Mouton, 1963), pp. 52–77

Broome, Matthew R., 'Suffering and Eternal Recurrence of the Same: The Neuroscience, Psychopathology, and Philosophy of Time', *Philosophy, Psychiatry, & Psychology*, vol. 12, no. 3 (2005), pp. 187–94

Bruner, Jerome, 'Life as Narrative', *Social Research*, vol. 71, no. 3 (2004), pp. 691–710

——, *Acts of Meaning: Four Lectures on Mind and Culture* (Cambridge, MA: Harvard University Press, 1992)

Butler, Robert N., *The Longevity Revolution: The Benefits and Challenges of Living a Long Life* (New York: PublicAffairs, 2008)

Carey, Benedict, 'For the Brain, Remembering Is Like Reliving', *New York Times*, 5 September 2008

Changeux, Jean-Pierre, *The Physiology of the Truth: Neuroscience and Human Knowledge*, trans. M. B. DeBevoise (Cambridge, MA: Harvard University Press, 2002)

Connolly, William E., *Neuropolitics: Thinking, Culture, Speed* (Minneapolis: University of Minnesota Press, 2002)

Conrad, Peter, *Modern Times, Modern Places: Life and Art in the Twentieth Century* (London: Thames and Hudson, 1998)

Csikszentmihalyi, Mihaly, *Beyond Boredom and Anxiety: Experiencing Flow in Work and Games* (San Francisco: Jossey-Bass, 1975)

—, *Flow: The Psychology of Happiness* (London: Rider, 1992)

Damasio, Antonio, *The Feeling of What Happens: Body & Emotion in the Making of Consciousness* (London: Heinemann, 2000)

—, *Looking for Spinoza* (London: Vintage, 2004)

Dawson, Kim A., 'Temporal Organization of the Brain: Neurocognitive Mechanisms and Clinical Implications', *Brain and Cognition*, vol. 54 (2004), pp. 75–94

Davies, Paul, *About Time: Einstein's Unfinished Revolution* (London: Viking, 1995)

Dodgshon, Robert A., *Society in Time and Space: A Geographical Perspective on Change* (Cambridge: Cambridge University Press, 1998)

Edelman, Gerald M., *Wider than the Sky: A Revolutionary View of Consciousness* (London: Penguin, 2005)

Ellis, Mary Lynne, *Time in Practice: Analytical Perspectives on the Times of Our Lives* (London: Karnac, 2008)

Erikson, Thomas Hylland, *The Tyranny of the Moment: Fast and Slow Time in the Information Age* (London: Pluto, 2001)

Firan, Carmen, 'Inside and Beyond Words', *Aspasia*, vol. 2, no. 1 (2008), pp. 195–200

Foster, Russell G. and Katharina Wulff, 'The Rhythms of Rest and Excess', *Nature Reviews Neuroscience*, vol. 6 (May 2005), pp. 407–14

Fraser, J. T., *Time: The Familiar Stranger* (Amherst: University of Massachusetts Press, 1987)

Freud, Sigmund, *An Outline of Psychoanalysis*, trans. James Strachey (New York: W. W. Norton, 1949)

Fuchs, Thomas, 'Fragmented Selves: Temporality and Identity in Borderline Personality Disorder', *Psychopathology*, vol. 40, no. 6 (2007), pp. 379–87

Galison, Peter, *Einstein's Clocks and Poincaré's Maps* (London: Sceptre, 2003)

Geertz, Clifford, *The Interpretation of Cultures* (New York: Basic Books, 1997)

Gleick, James, *Faster: The Acceleration of Just About Everything* (New York: Vintage, 2000)

—, *Faster: Our Race Against Time* (London: Abacus, 2005)

Glicksohn, Joseph, 'Temporal Cognition and the Phenomenology of Time: A Multiplicative Function for Apparent Duration', *Consciousness and Cognition*, vol. 10, no. 1 (2001), pp. 1–25

Goodin, Robert E. et al., *Discretionary Time: A New Measure of Freedom* (Cambridge: Cambridge University Press, 2008)

Green, André, *Time in Psychoanalysis: Some Contradictory Aspects*, trans. Andrew Weller (London: Free Association Books, 2002)

Greer, Mark, 'Strengthen Your Brain by Resting It', *APA: Monitor on Psychology* vol. 35, no. 7 (2004) http://www.apa.org/monitor/julaug04/strengthen.html [accessed 5 Feb. 2009]

Hartocollis, Peter, *Time and Timelessness, or The Varieties of Temporal Experience* (New York: International Universities Press, 1983)

Hawking, Stephen, *A Brief History of Time: From the Big Bang to Black Holes* (Toronto: Bantam, 1995)

Heims, Steve J. et al., '"In the River of Consciousness": An Exchange', *New York Review of Books*, 8 April 2004, http://www.nybooks.com/articles/17030 [accessed 5 Feb. 2009]

Hochachka, Peter W. and Michael Guppy, *Metabolic Arrest and the Control of Biological Time* (Cambridge, MA: Harvard University Press, 1987)

Hochschild, Arlie Russell, *The Time Bind: When Work Becomes Home and Home Becomes Work* (New York: Henry Holt, 2001)

Honoré, Carl, *In Praise of Slow: How a Worldwide Movement is Challenging the Cult of Speed* (London: Orion, 2005)

Husserl, Edmund, *On the Phenomenology of the Consciousness of Internal Time, 1893–1917*, trans. John Barnett Brough (Dordrecht: Kluwer Academic, 1992)

Ivry, Richard B. and Rebecca M. C. Spencer, 'The Neural Representation of Time', *Current Opinion in Neurobiology*, vol. 14 (2004), pp. 225–32

Jaffe, Robert L., 'As Time Goes By', *Natural History Magazine*, October 2006, http://www.naturalhistorymag.com/1006/1006_feature2.html [accessed 5 Feb. 2009]

Johnston, Adrian, *Time Driven: Metapsychology and the Splitting of the Drive* (Evanston, IL: Northwestern University Press, 2005)

Kafka, Franz, *The Great Wall of China*, trans. Willa Muir and Edwin Muir (New York: Schoken, 1946)

Kandel, Eric, *In Search of Memory: The Emergence of a New Science of Mind* (New York: W. W. Norton, 2007)

——, *Psychology, Psychoanalysis and the New Biology of Mind* (Arlington, VA: American Psychiatric Publishing, 2005)

Kapuscinski, Ryszard, *The Shadow of the Sun,* trans. Klara Glowczewska (New York: Vintage, 2002)

Kirn, Walter, 'The Autumn of the Multitaskers', *Atlantic Monthly*, November 2007, pp. 66–75

Klarsfeld, André, and Frédéric Revah, *The Biology of Death: Origins of Mortality*, trans. Lydia Brady (New York: Cornell University Press, 2003)

Klein, Stefan, *Time: A User's Guide* (London: Penguin, 2008)

Kreitzman, Leon, *The 24 Hour Society* (London: Profile, 1999)

Kundera, Milan and Linda Asher, *Slowness* (London: Faber and Faber, 1996)

Lacan, Jacques, *Écrits: A Selection*, trans. Alan Sheridan (London: Routledge, 1990)

Ledoux, Joseph, *The Emotional Brain: The Mysterious Underpinnings of Emotional Life* (New York: Simon and Schuster, 1996)

Lefebvre, Henri, *Rhythmanalysis: Space, Time and Everyday Life*, trans. Stuart Elden and Gerald Moore (London: Continuum, 2004)

Levine, Robert V., *A Geography of Time: The Temporal Misadventures of a Social Psychologist, or How Every Culture Keeps Time Just a Little Bit Differently* (New York: Basic Books, 1998)

Lifton, Robert Jay, *The Protean Self: Human Resilience in an Age of Fragmentation* (Chicago: University of Chicago Press, 1993)

Lightman, Alan, *The Diagnosis: A Novel* (New York: Vintage, 2002)

Linder, Staffan Burenstam, *The Harried Leisure Class* (New York: Columbia University Press, 1971)

Meissner, William, *Time, Self, and Psychoanalysis* (Lanham, MD: Jason Aronson, 2007)

Merleau-Ponty, Maurice, *Phenomenology of Perception*, trans. Colin Smith (London: Routledge, 2002)

Mooallem, Jon, 'The Sleep-Industrial Complex', *New York Times*, 18 November 2007, p. 56

Nabokov, Vladimir, *Speak, Memory: An Autobiography Revisited* (London: Penguin, 2000)

Nagel, Thomas, 'What Is It Like to Be a Bat?', *Philosophical Review*, vol. 83, no. 4 (1974), pp. 435–50

Nobre, Anna Christina and Jill O'Reilly, 'Time Is of the Essence', *TRENDS in Cognitive Sciences*, vol. 8, no. 9 (2004), pp. 387–9

Pachoud, Bernard, 'The Teleological Dimension of Perceptual and Motor Intentionality', in Jean Petitot et al., eds., *Naturalizing Phenomenology: Issues in Contemporary Phenomenology and Cognitive Science* (Stanford, CA: Stanford University Press, 1999), pp. 196–219

Perelberg, Rosine Jozef, *Time, Space, and Phantasy* (London: Routledge, 2008)

Phillips, Adam, *Darwin's Worms* (London: Faber and Faber, 1999)

Piaget, Jean, *The Child's Conception of Time* (New York: Routledge, 2006)

Posner, Michael, et al., 'Training, Maturation, and Genetic Influences on the Development of Executive Attention', *Proceedings of the National Academy of Sciences*, vol. 102, no. 41 (2005), pp. 14931–6

Proust, Marcel, *Remembrance of Things Past*, 3 vols, trans. C. K. Scott Moncrieff and Terence Kilmartin (New York: Vintage, 1982)

Restak, Richard, *The New Brain: How the Modern Age Is Rewiring Your Mind* (New York: Rodale, 2004)

Richtel, Matt, 'Lost in E-Mail, Tech Firms Face Self Made Beast', *New York Times*, 14 June 2008

Ricoeur, Paul, *Oneself as Another*, trans. Kathleen Blamey (Chicago: University of Chicago Press, 1992)

Rifkin, Jeremy, *Time Wars: The Primary Conflict in Human History* (New York: Henry Holt, 1987)

Rosenfield, Israel and Edward Ziff, 'How the Mind Works: Revelations', *New York Review of Books*, 26 June 2008, pp. 62–5

Russell, Bertrand, *In Praise of Idleness: And Other Essays* (New York: Routledge, 2004)

Sabbadini, Andrea, 'Boundaries of Timelessness: Some Thoughts About the Temporal Dimension of the Psychoanalytic Space', *International Journal of Psycho-Analysis*, vol. 70, no. 1 (1989), pp. 305–13

Sacks, Oliver, 'In the River of Consciousness', *New York Review of Books*, 15 January 2004

——, *The Man Who Mistook His Wife for a Hat: And Other Clinical Tales* (New York: Touchstone, 1998)

Savigar, Tom, *Born Clicking: Are Kids Smarter than Adults?* (London: Institute of Contemporary Arts, 2002)

Scheuerman, William E., *Liberal Democracy and the Acceleration of Time* (Baltimore, MD: Johns Hopkins University Press, 2004)

Smith, Jonathan D. H., 'Time in Biology and Physics', in R. Buccheri, M. Saniga, and W. M. Stuckey, eds., *The Nature of Time: Geometry, Physics and Perception* (Dordrecht: Kluwer, 2003), pp. 145–52; available at http://www.chronos.msu.ru/EREPORTS/smith_time.pdf [accessed 5 Feb. 2009]

Squire, Larry R. et al., eds., *Fundamental Neuroscience* (London: Elsevier Science, 2002)

Steinhauer, Jennifer and Laura M. Holson, 'As Text Messages Fly, Danger Lurks', *New York Times*, 20 September 2008

Thompson, Clive, 'Brave New World of Digital Intimacy', *New York Times*, 7 September 2008

Tipler, Frank J., *The Physics of Immortality* (New York: Doubleday, 1994)

Tomlinson, John, *The Culture of Speed: The Coming of Immediacy* (London: Sage, 2007)

Varela, Francisco J., 'The Specious Present: A Neurophenomenology of Time Consciousness', in Jean Petitot et al., eds., *Naturalizing Phenomenology: Issues in Contemporary Phenomenology and Cognitive Science* (Stanford, CA: Stanford University Press, 1999), pp. 226–314

Veen, Wim and Ben Vrakking, *Homo Zappiens: Growing up in a Digital Age* (London: Continuum, 2006)

Virilio, Paul, *The Information Bomb*, trans. Chris Turner (London: Verso, 2000)

—, *The Lost Dimension: Zero Gravity*, trans. Daniel Moshenberg (New York: Semiotext(e), 1991)

Williams, Paul, 'Making Time, Killing Time', in Rosine Jozef Perelberg, ed., *Time and Memory: The Power of the Repetition Compulsion* (London: Karnac, 2007), pp. 47–64

Willock, Brent et al., eds., *On Deaths and Endings: Psychoanalysts' Reflections on Finality, Transformations and New Beginnings* (New York: Routledge, 2007)

Winterson, Jeanette, *The World and Other Places* (London: Vintage, 2000)

Wolin, Sheldon S., 'What Time Is It?', *Theory and Event*, vol. 1, no. 1 (1997), pp. 1–10

Wood, David, ed., *On Paul Ricoeur: Narrative and Interpretation*, Warwick Studies in Philosophy and Literature (London: Routledge, 1992)

Zimmer, Carl, 'Time in the Animal Mind', *New York Times*, 3 April 2007

Zuckerman, Marvin, 'The Neurobiology of Impulsive Sensation Seeking: Genetics, Brain Physiology, Biochemistry and Neurology', in Con Stough, ed., *Neurobiology of Exceptionality* (New York: Kluwer Academic/Plenum Publishers, 2005), pp. 31–52

ACKNOWLEDGEMENTS

My great thanks go to Lisa Appignanesi, for encouraging me to follow my thematic impulses, and for her editorial wisdom and tact; to Andrew Franklin and Frances Coady, for trusting me to take on an unexpected subject, and their truly inspiriting support; to Trevor Horwood and David Rogers for their discerning responses to the manuscript; and to the staff members at both Profile and Picador who have worked with unfailing energy and patience on this project.

INDEX